Undergraduate Lecture Notes in Physics

For further volumes:
http://www.springer.com/series/8917

Undergraduate Lecture Notes in Physics (ULNP) publishes authoritative texts covering topics throughout pure and applied physics. Each title in the series is suitable as a basis for undergraduate instruction, typically containing practice problems, worked examples, chapter summaries, and suggestions for further reading.

ULNP titles must provide at least one of the following:

- An exceptionally clear and concise treatment of a standard undergraduate subject.
- A solid undergraduate-level introduction to a graduate, advanced, or non-standard subject.
- A novel perspective or an unusual approach to teaching a subject.

ULNP especially encourages new, original, and idiosyncratic approaches to physics teaching at the undergraduate level.

The purpose of ULNP is to provide intriguing, absorbing books that will continue to be the reader's preferred reference throughout their academic career.

Ganesan Srinivasan

What are the Stars?

Ganesan Srinivasan
Bangalore
India

Originally published by Universities Press (India) Private Limited 2011, ISBN 9788173717413

ISSN 2192-4791 ISSN 2192-4805 (electronic)
ISBN 978-3-642-45301-4 ISBN 978-3-642-45302-1 (eBook)
DOI 10.1007/978-3-642-45302-1
Springer Heidelberg New York Dordrecht London

Library of Congress Control Number: 2013956738

Printed on acid-free paper

Springer is part of Springer Science+Business Media (www.springer.com)

Foreword to the Edition Published by Universities Press

Lord Martin Rees
Professor of Cosmology and Astrophysics
Astronomer Royal
Master of Trinity College, Cambridge
Past President, Royal Society

If you chose 10,000 people at random, 9,999 would have something in common—their business and their interests would lie on or near the Earth's surface. The other would be an astronomer. I'm lucky to be one of this strange breed—as is Dr. G. Srinivasan, the author of this series of monographs entitled *The Present Revolution in Astronomy*. But astronomy isn't just for astronomers. Its findings are fascinating, and it is as important to understand the cosmos as it is to appreciate the rest of nature. The entire cosmos is part of our environment. Indeed the dark night sky is one feature that's been essentially unchanged throughout all human history, shared by all cultures—though it has been interpreted in many different ways.

Astronomers are the heirs to a long tradition. Astronomy is the oldest science—except perhaps for medicine. Its origins lie in the need to establish a calendar, to measure time, and to interpret the patterns and regularities seen in the sky. Our knowledge is now advancing faster than ever before—thanks to powerful telescopes, and probes that travel to other planets. A wide public has shared the excitement of this vicarious exploration.

We can't send actual probes beyond our Solar System, but with our telescopes, we can study stars in detail. In the last decade we have learnt something that's made the night sky far more interesting. Stars aren't mere twinkling 'points of light'. They're orbited by retinues of planets, just like the Sun is. Some of these planets may be like the Earth—but whether there is life on any of them is a question that challenges future generations of scientists.

We have come to realise the immense scale of the universe—in both space and time. We live in a galaxy containing more than a hundred billion stars; but this galaxy is itself just one of a hundred billion visible with modern telescopes. By looking far away in space, we can penetrate far back in time, because the light from distant objects took a long time to reach us. Astronomers have an advantage over geologists and fossil hunters: they can actually observe the past, and trace cosmic history right back to the formation of the first stars and galaxies. Indeed there is compelling evidence that our universe is the expanding aftermath of a "big bang" nearly 14 billion years ago.

We have learnt one crucial thing about the universe: it is governed by physical laws that we can understand, and these laws seem to be the same everywhere. By analysing the light from a distant galaxy, we can infer that the atoms it's made of behave just like those we study in the laboratory. It's because of this uniformity that we can understand the structure of stars and their life cycles, and how, from simple beginnings, stars, galaxies and planets emerged to form the complex structured cosmos of which we are a part.

The cosmos is a unity. There are links between the very small—the microworld of atoms—and the very large—stars and galaxies. Stars form, evolve and die (sometimes explosively). They are powered by nuclear fusion—a controlled version of what happens in a hydrogen bomb. Over their lifetime, this process generates, from pristine hydrogen, atoms of carbon, oxygen and iron. All the atoms on Earth, and in our bodies are the ashes from long-dead stars. We are the 'nuclear waste' from the fusion power that makes stars shine. Fully to understand ourselves and our origins, we must understand not only Darwinian evolution, but also the atoms all life is made of, and the stars that made those atoms. This wonderful story should be part of everyone's education.

But there is another reason for studying astronomy. It allows us to probe the laws of nature under far more extreme temperatures, pressures and energies than can be achieved in laboratories here on Earth. It also allows us to study the fundamental force of gravity, and how it relates to the nature of space and time.

This is undoubtedly the Golden Age of astronomy. With the advent of the space age, new windows to the Universe have been opened. With giant observatories orbiting high above the Earth's atmosphere, one can now explore the Universe at a wide range of wavelengths: radio waves, millimetre waves, infrared radiation, visible radiation, ultraviolet radiation, X-rays and gamma rays. This has enabled astronomers to make unprecedented progress pertaining to a variety of questions: the nature of the stars and their life history; the formation of planets; the birth and death of the stars; the graveyard of stars—white dwarfs, neutron stars and black holes; galaxies; quasars; and the Universe at large.

This series of monographs entitled *'The Present Revolution in Astronomy'* is very timely for it aims to survey the contemporary scene at an introductory level. Dr. G. Srinivasan, the author of this series of books, is an internationally acclaimed leader in this enterprise. In particular, he has studied neutron stars, which manifest an astonishing range of 'extreme' physics. Readers of these splendid and accessible books will find Dr. Srinivasan to be a clear and enthusiastic guide to the wonders and mysteries of the cosmos. We should all be grateful to him.

Cambridge Martin J. Rees

Preface to the Edition Published by Universities Press

The year 2009 was celebrated as the *International Year of Astronomy*. This was to commemorate the 400th anniversary of Galileo's pioneering observations with a telescope, observations that revolutionized man's perception of the heavenly bodies.

Four centuries later, we are in the midst of another golden era in astronomy. The advent of the space age has opened new windows to the Universe, resulting in spectacular discoveries and unprecedented progress in our understanding of the nature of celestial objects. At the same time, many new and outstanding questions have emerged. Indeed, there are clear indications that the resolution of some of these puzzles may require a major revision of fundamental physics itself. A deep connection between the *microcosm* and the *macrocosm* is becoming apparent.

This series of monographs entitled *The Present Revolution in Astronomy* is intended to convey the excitement of contemporary astronomy. The inspiration for writing these monographs was the enthusiastic response of the students who attended an intercollegiate course I taught for 5 years at St. Joseph's College in Bangalore. This course was not part of the regular academic curriculum, and was open to interested students and teachers from all the colleges in the city. Interestingly, more than half the students in each batch were students of engineering, rather than pure science. And yet, they were fascinated by the lure of astronomy. Although the underlying theme of the course was *The Present Revolution in Astronomy*, my idea was to use astronomy as a *Trojan horse* to get the young students excited about the challenges that await them in the world of physics/astronomy, engineering and technology. It was the unanimous view of these students that I should develop these lectures into a series of books.

There is a second reason why I thought it would be worthwhile to write these books. Historically, astronomy has always had a great appeal among the general public. It is even more so today. The commissioning of new telescopes, and the discoveries made with them receive wide publicity in the print as well as the electronic media. Space Agencies like NASA, as well as leading astronomical observatories, have impressive Public Outreach programmes. And yet, here in India, hardly any of the universities offer astronomy as one of the subjects in the undergraduate curriculum. As a result of the lack of familiarity with the subject, very few students opt for a career in astronomy even though there are several truly world class observing facilities in India. This series of books is intended to partly remedy this lacuna.

Now a few words about the scope of these monographs and the style in which they are written. My primary objective is to introduce the reader, young and not so young (!), to the presently unfolding revolution in astronomy. We shall discuss the recent developments concerning a wide variety of topics: *the nature of the stars and their life history; the birth and death of the stars; the graveyard of stars—white dwarfs, neutron stars and black holes; galaxies; quasars; and the Universe at large.*

The monographs are not intended to be 'textbooks' in astronomy. Textbooks have to develop the subject in a pedagogical manner, dwell on the experimental methods and phenomenology, develop the mathematical aspects of the theory in a systematic manner, include problems and exercises, etc. While all these are needed to learn a subject seriously, conventional textbooks have a serious handicap. Introductory books 'begin at the beginning' and seldom convey the excitement surrounding contemporary developments. They tend to focus on questions that have been resolved, rather than highlight what is not known. In contrast, this series of books is intended to serve a different purpose. I hope they will give the reader an introduction to the recent developments, as well as highlight the outstanding and unsolved questions. I believe that a young reader would be more interested in the unsolved puzzles, for that is where the challenges lie.

The books have a very different flavour compared to the traditional astronomy books. For example, they do not discuss topics such as measurement of distances to celestial objects, determination of their masses, luminosities, etc. Nor do they dwell on coordinate systems to define their positions in the sky, the classification of their spectra, etc. While all these are 'bread and butter' issues, it is my view that a reader would learn these topics at a later stage in the normal course if he or she decides to become a practising astronomer. The emphasis in this series of monographs will be on physics, and for the following reason.

Among the many great discoveries made by Isaac Newton, perhaps the most profound was his assertion that *the Laws of Nature have universal validity*. In other words, the laws of physics that govern phenomena on Earth apply everywhere in the Universe. Today, we take this assertion by Newton as an axiom. Indeed, during the past couple of centuries, several seminal inputs to laboratory physics have come from astronomical observations. The discovery of the law of gravitation, emission and absorption lines in the spectrum of the atoms, the discovery of Helium, the first verification of the predictions of the Special Theory of Relativity and the General Theory of Relativity are some of the more important examples. This is not surprising. The range of densities, temperatures and pressure that are obtained in celestial bodies are staggering compared to what one encounters on earth. For example, the densities range from 1 atom/cm^3 to 10^{37} atoms/cm^3, and the temperatures range from 3 kelvin to 100 million kelvin—conditions that are hard for us to comprehend. Consequently, one encounters many new and exotic physical phenomena in celestial objects. Indeed, a few decades ago one would have said that *'Astronomy is the home of Physics'*. Today, however, it would be more appropriate to say that *'Physics is the home of astronomy'*. We shall see the reason for this paradigm shift as we progress in this series. Therefore, we shall concentrate on the physics of the celestial bodies—their nature, their stability, their central engines, their radiation mechanisms, etc.

Having stated the objective of this series of books, I must add that I do not assume any astronomical background from the reader. A knowledge of physics at, say, the *Halliday and Resnick* level would be quite adequate to get started. We shall develop the rest of the background as we go along. To meet the stated objectives, I shall often be required to sacrifice rigour in the arguments in favour of simple analogies and qualitative arguments. And I shall do so without any apologies! I shall consider my efforts worthwhile if these books manage to convey the excitement of contemporary astronomy. As for the younger readers, I do hope that these books will arouse their interest sufficiently enough for them to want to pursue the topics further by going to more learned books.

When I was young, I had the pleasure and privilege to read the marvellous books by great masters like Sir Arthur Eddington, Sir James Jeans and George Gamow, books in which they explained the developments in physics and astronomy in the early part of the last century. There are several recent books, written in a similar vein, by leading physicists and astrophysicists, of the present epoch. And then there is the 'Internet'! This series of monographs represents my very humble efforts in the same spirit.

This Volume

This book begins with an overview of the present revolution in astronomy. It should give you a feel for the topics we shall be discussing in this series of monographs. The rest of this book is devoted to a discussion of the nature of the stars, their stability and the source of the energy they radiate. This is where the subject of astrophysics began. The story begins in the early decades of the nineteenth century. Although the foundations of the subject were laid by 1870, the edifice was built only in the 1920s. Much of our understanding of the nature of the stars dates back to that period. As I have mentioned above, this series of books is not just about what has been well understood—it is about the current excitement in astrophysics as well. Interestingly, the definitive proof of many of the prescient conjectures made in the 1920s and 1930s came less than ten years ago. The last two chapters of this book are devoted to these recent developments.

Acknowledgments

The idea of this series was first suggested by the students who attended the intercollegiate course on astronomy and astrophysics that I taught for a number of years at St. Joseph's College in Bangalore. This suggestion was strongly endorsed by Dr. P. Sreekumar of the ISRO Satellite Centre. The enthusiastic response of the student community to the series of books entitled *Vignettes in Physics*, written by Dr. G. Venkataraman, as well as Venkataraman's eloquent and sustained persuasion that I should write a similar series on contemporary astronomy, gave me the conviction I needed to undertake this task. A further impetus came when the Jawaharlal Nehru Memorial Fund bestowed on me the *Jawaharlal Nehru Fellowship* in 2007 to get started on this project. In 2009, the Nehru Centre in Mumbai was very kind to give me a Fellowship for two years to continue with the project. I am very grateful for both these Fellowships. I started out as a condensed matter physicist, but later wandered into astronomy! My first introduction to astronomy came from my father at an early age. The inspiration to pursue it and the attempt to popularize it, came first from my illustrious teacher Subrahmanyan Chandrasekhar (at the University of Chicago), and later from Professors Martin Rees (Cambridge University), Ed van den Heuvel (University of Amsterdam) and V. Radhakrishnan (at the Raman Research Institute, Bangalore). I am most grateful to them for having inspired me. I would like to express my special thanks to NASA, ESA, and the international astronomical fraternity for the many wonderful images reproduced in these volumes.

G. Srinivasan

Contents

About the Author

Dr. G. Srinivasan began his career as a solid state physicist and later switched to astrophysics. After his Ph.D. at the University of Chicago, he worked at the IBM Research Laboratory, Zurich, Switzerland, Chalmers University of Technology, Goteborg, Sweden, Cavendish Laboratory, University of Cambridge and Raman Research Institute, Bangalore. He is a Past President of the Astronomical Society of India as well as the Division of Space and High Energy Astrophysics of the International Astronomical Union. He is a Fellow of the Indian Academy of Sciences and a former Jawaharlal Nehru Fellow.

The Present Revolution in Astronomy: An Overview

The year 2009 was celebrated as the *International Year of Astronomy*. This was to commemorate the four-hundredth anniversary of Galileo's pioneering observations with a telescope, observations that revolutionized man's perception of the heavenly bodies.

Four centuries later, we are in the midst of another golden era in astronomy. The advent of the space age has opened new windows to the Universe, resulting in spectacular discoveries and unprecedented progress in our understanding of the nature of celestial objects. At the same time, many new and outstanding questions have emerged. Indeed, there are clear indications that the resolution of some of these puzzles may require a major revision of fundamental physics itself.

This series of monographs is intended to convey the excitement of contemporary astronomy, and serve as an introduction to the present revolution in astronomy. The purpose of this overview is to the set the scene, so to speak. It is not my intention to explain anything systematically in this broad-brush outline. It is intended to be more like a *trailer* of a soon-to-be released movie. It is my hope that this introduction will whet your appetite sufficiently to make you look forward to the other volumes of this series to learn more about the revolution that is currently unfolding.

The Dawn of the Twentieth Century

What does one mean by a revolution in science? Every now and then, questions which have remained meaningless, or considered frivolous, suddenly acquire meaning within the premise of science. When this happens, a scientific revolution begins. Let us recall a couple of examples. When Isaac Newton stated that *natural phenomena should be understood in terms of underlying physical laws*, it was a revolutionary statement; no one had asserted this before. Far more profound was Newton's assertion that *the laws of nature are of Universal validity*; that is, the same laws apply everywhere in the Universe.

The revolution that was unfolding at the dawn of the twentieth century concerned the nature of stars. According to the positivist philosophers who greatly

Fig. 1 The great spiral galaxy M31 in the constellation Andromeda (from the Wikimedia commons, with the kind permission of the author, John Lanoue). Countless number of such spiral *nebulae* were thought to be part of own Galaxy till Edwin Hubble established that this *nebula* was at a distance of nearly three million light years. Since our Galaxy is only a hundred thousand light years across, M31 could not be in our Galaxy. It had to be a galaxy in its own right!

influenced European thinking in the eighteenth and nineteenth century, *it was in the nature of things that we shall never know what the stars are*. The discovery of the dark lines in the spectrum of the Sun and the stars by Fraunhofer in 1817, and their subsequent explanation by Kirchoff, Bunsen and others, proved the philosophers wrong. It was clear that at least the outer layers of the Sun was gaseous and made of the same atoms that we find on earth. The subject of astrophysics was born. By the 1930s one had understood a great deal about what are the stars, and why are they as they are.

The great new question that arose at the beginning of the twentieth century concerned the size and the nature of the astronomical Universe. A much debated question at that time was whether the Universe was synonymous with our own Milky Way galaxy. The perception at that time was that our galaxy was rather small—roughly 20,000 light years across—and the Sun was at the centre of the Galaxy. In 1923 the astronomer Edwin Hubble was able to estimate the distance to the great Spiral nebula in the constellation Andromeda (Fig. 1).

It became clear that the Andromeda nebula was at a distance of approximately three million light years. This established beyond all doubt that it was a galaxy in its own right. Today, we know that there are more than 350 *billion* giant galaxies in the Universe, each containing about a 100 billion stars. Soon, Hubble went on to demonstrate that these countless galaxies were not the building blocks of the Universe. The building blocks of the Universe were *clusters of galaxies*, with each cluster containing several-hundred to several-thousand galaxies. The number of clusters of galaxies is now estimated to be more than 25 *billion*. These clusters of galaxies are the building blocks of the Universe, just as molecules are the building blocks of a gas. Thanks to these two remarkable discoveries by Hubble, the Universe became much bigger than what one had envisaged at the beginning of the twentieth century (Fig. 2).

Fig. 2 The building blocks of the Universe: Clusters of galaxies with several-hundred to thousand of galaxies are the building blocks of the Universe. This photograph was taken with the Hubble Space Telescope. [Courtesy of NASA]

In 1929, Hubble made an even more astonishing discovery. He discovered that the clusters of galaxies were systematically receding from us; this was true no matter which cluster one looked at. In fact, the velocity of recession of the cluster was linearly dependent on its distance from us. That is, a cluster which is twice as far away is receding from us twice as fast. What is one to make of this? If taken literally, this would imply that in the distant past the clusters of galaxies must have been closer together (Fig. 3).

The Belgian physicist Georges Lemaître went one giant step further. He suggested that the recession of the clusters discovered by Hubble might actually be the local signature of the expansion of the Universe as a whole. Indeed, Lemaître proposed that the Universe must have once been a *primeval atom!* In the 1940s, the brilliant Russian physicist George Gamow took the bold step of taking all this seriously and making far reaching predictions. One of these was that if the early Universe was dense and hot, it would be an excellent laboratory to synthesize the elements by fusing hydrogen nuclei. Such a suggestion had been made earlier by Sir Arthur Eddington in the context of the stars. The other important prediction was that the Universe today must contain the relic of the primeval radiation and that the temperature of this all pervading radiation must be a few degrees kelvin.

Not many took all this seriously, although this was the only scenario that could explain the abundance of the light elements in the Universe such as deuterium, helium and lithium. The reason for this scepticism was that Gamow's conjecture represented an extrapolation into the unknown. For one thing, he (like Lemaître) generalized the fairly *local phenomenon* of the recession of the clusters of galaxies that Hubble had detected to imply a Universal expansion. He then extrapolated backwards in time into a domain where one didn't know the laws of physics. The question of the origin of the Universe could not yet be posed rigorously within the premise of the existing theories.

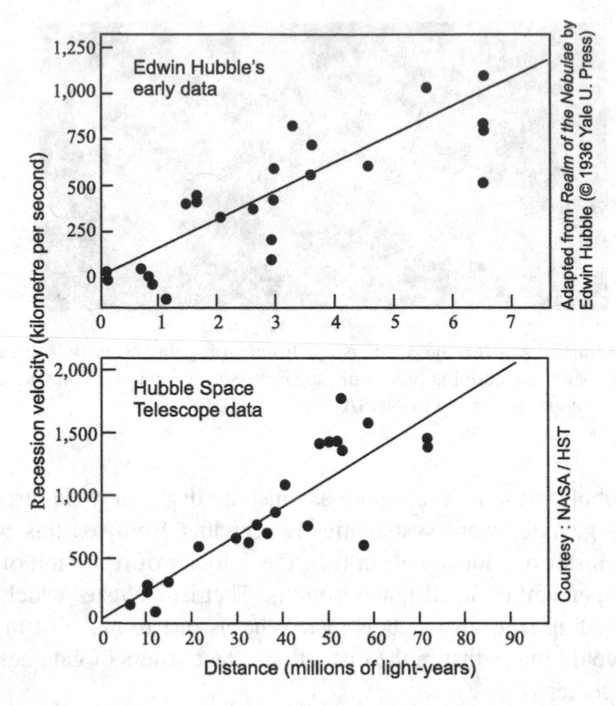

Fig. 3 The recession of the clusters of galaxies and Hubble's Law. The *top panel* shows Hubble's original data, while the *lower panel* shows more recent data. Plotted on the y-axis is the recession velocity in kilometres per second. The distance to the clusters, in millions of light years, is plotted on the x-axis. Notice the different scales on the x-axis in the two plots. The solid line indicates the linear relationship between the two. This is Hubble's law

The Giant Strides of the Twentieth Century

By the time the twentieth century drew to a close, great progress had been made in our understanding of the nature of celestial objects. Let us briefly review some of it before coming to where the action is today.

What are the Stars?

As already mentioned, Fraunhofer's discovery of dark lines in the spectrum of the Sun enabled the physicists to conclude that the Sun's outer layers were gaseous. By 1870, the Sun and the stars had been modelled as gaseous spheres, held together by their own gravity. The outstanding question at the turn of the nineteenth century was the following: *what is the source of energy that makes the stars shine?* In 1920, Sir Arthur Eddington at Cambridge University in England made the extraordinary suggestion that the source of energy was the transmutation of hydrogen into helium at the centre of the stars. He went on to construct a detailed theory of the stars based upon the simple principle that the inward-directed force due to self gravity was balanced by the combined pressure of the gas and radiation, both of which are directed outwards. Despite its simplicity, many predictions of this theory were in remarkable agreement with observations; yet, a detailed understanding of the source of energy radiated by the stars had to wait till the emerging discipline of Nuclear Physics had come of age.

In 1938, Hans Bethe worked out all the details of the fusion reactions. His results for the energy released per second by the Sun by fusing hydrogen into helium agreed extremely well with the energy radiated by the Sun per unit time, but there was no direct evidence of such fusion reactions taking place in the centre of the Sun. Such evidence could only be provided by the other by products of the reactions. It was therefore imperative to detect the *neutrinos* produced when hydrogen is transmuted into helium. If Eddington and Bethe were correct, then the Sun should be emitting approximately 10^{38} neutrinos every second. Since a fair number of these neutrinos would reach the earth, it would appear to be a simple task to detect them. Unfortunately, neutrinos interact incredibly weakly with matter. But physicists were undeterred. The neutrinos from the Sun were finally detected in 1968. However, there was an irritating problem. The number of neutrinos detected was only one-third of what the theory had predicted. This puzzle challenged the physicists for three decades and was solved only in 2001. So the mystery of the source of energy radiated by the stars was finally solved.

Can Stars Find Peace?

A major problem arose in 1925 when an extraordinary star was discovered. Although it was as massive as the Sun, it was only as big as the Earth! This meant that the average density of the star was about 10^6 g cm^{-3} (the mean density of the Sun is only a little more than that of water). The difficulty posed by such a dense star was the following. What will happen to such a star when nuclear energy generation at its centre stops? Since the star will no longer produce heat, there will be no force to oppose gravity, and the star will have no option but to collapse under its own gravity. What will be the end state of such a star?

Surprisingly, the resolution of this problem came from the newly emerging *Quantum Physics*. In 1926, R. H. Fowler at Cambridge University argued that the star will collapse and collapse till it reaches a density where a new quantum mechanical force provides support against gravity, and the star will at last find peace. This quantum pressure is due to the electrons, and arises due to the combined effect of Heisenberg's *Principle of Uncertainty* and Pauli's *Exclusion Principle*. Soon, Subrahmanyan Chandrasekhar, a student at Presidency College in Madras developed Fowler's idea into a proper theory of such quantum stars, which had come to be known as *white dwarfs*. And thus it appeared to be established that *all stars* will ultimately find ultimate peace as white dwarfs.

This sense of security was shattered in 1930, when Chandrasekhar discovered that such white dwarfs cannot be more massive than 1.4 times the mass of the Sun. This raised an embarrassing question. What will happen to stars that are more massive than $1.4 M_{\odot}$? Chandrasekhar's discovery implied that all such stars will collapse indefinitely and become singularities—regions of spacetime of zero volume and infinite density.

The Discovery of Radio Astronomy

The year 1928 was a momentous one in the history of astronomy. Karl Jansky, working at the Bell Telephone Laboratory in New Jersey, USA, discovered radio waves emitted by celestial objects. The pursuit of this monumental discovery was impeded by the outbreak of World War II. After the war ended, scientists who were involved in the development of the *Radar* turned their attention to the pursuit of astronomy at radio wavelengths. Large telescopes were built in a quest for higher angular resolution and higher sensitivity. This led to many major discoveries that revolutionized astronomy. These included the discovery of quasars and neutron stars. Another discovery in the realm of radio astronomy that had a great impact was the discovery of the 21-*cm radiation* from neutral hydrogen atoms.

By the turn of the twentieth century, giant radio telescopes had been built in many countries, operating at a wide range of wavelengths, ranging from 10 m to a few millimetres. One of them, the Giant Metrewave Radio Telescope (GMRT), the largest radio telescope in the world operating at a wavelength of approximately 1 metre, is located near Pune in India. Using these telescopes, astronomers have been able to probe the distant corners of the Universe, and study its structure and dynamics (Fig. 4).

New Windows to the Universe

More windows to the Universe were opened with the advent of the space age. By the mid 1960s, astrophysics had become a jig-saw puzzle. To understand the nature of a celestial object one had to understand its structure and stability, its composition, the physical conditions, the radiation mechanism, etc. This required observations over a broad spectrum, ranging from radio waves to gamma rays, with each wavelength region providing one piece of the jig-saw puzzle. Unfortunately, Earth's atmosphere absorbs most of this radiation. Just to give you an idea, sub-millimetre-wave radiation is absorbed by the rotational levels of molecules in the air. Infrared radiation is absorbed by the vibrational levels of molecules. Ultraviolet radiation has the right energy to break up molecules, and

Fig. 4 One of the thirty telescopes of the GMRT (Giant Metrewave Radio Telescope) near Pune in India. This giant radio telescope was conceived and built by Govind Swarup and his colleagues at the Tata Institute of Fundamental Research in Mumbai

hence they disappear. X-ray photons lose their energy by ionizing atoms in the atmosphere. High-energy gamma rays disappear in the atmosphere by creating a *shower* of particle–antiparticle pairs. Therefore, to detect radiation at these wavelengths, one has to go above the Earth's atmosphere. The advent of the space age enabled one to do this. The first breakthroughs came from experiments on rockets. By mid 1970s, astronomers were able to launch giant telescopes orbiting the earth. The Einstein X-ray Observatory, the Infrared Astronomy Satellite (IRAS), the International Ultraviolet Explorer (IUE), the Compton Gamma Ray Observatory, the Hubble Space Telescope, the Cosmic Background Explorer are some early examples. In recent times, much more sensitive telescopes have been launched, and many more are in the pipeline. This is undoubtedly the era of exploring the Universe from outer space.

Exploding Stars

During the first millennium, the oriental astronomers, notably the Chinese, meticulously recorded the appearance and disappearance of new stars in the Sky. Sometimes, they could even be seen in daylight for many months! They called them *guest stars*. Astronomers in the Middle East and Europe also recorded many such guest stars during the second millennium; they called them *Novae* (new stars). In 1572 AD, the great Danish astronomer Tycho Brahe observed one such *Nova*. Being a great astronomer, he carefully recorded the rise and fall in the brightness of the new star by comparing it with the brightness of known stars in the sky. Johannes Kepler, too, had the privilege of observing a *Nova* in 1604.

One such guest star was seen in the great Andromeda nebula in 1885. The nature of these guest stars began to unravel itself soon after Hubble's discovery

Fig. 5 An optical image of the Crab Nebula, the expanding debris of the supernova explosion of 1054 AD. At the centre of this nebula is a rapidly spinning, strongly magnetized neutron star. [This remarkable image was taken with the Hubble Space Telescope. Courtesy of NASA]

that the Andromeda nebula was an external galaxy, nearly three million light years away. This enabled Fritz Zwicky, an incredibly creative Swiss theoretical physicist at the California Institute of Technology, to estimate for the first time the energy released by the guest star of 1885 in Andromeda; it was truly fantastic. Zwicky estimated that approximately 10^{52} erg of energy was released in a very short time. This is the total amount of energy that our Sun would radiate in its entire life time of several billion years. To put it differently, this is the total energy radiated in one second by all the stars in our galaxy, and there are approximately 10^{11} stars in a typical galaxy! Therefore, whatever they might be, these gust stars deserved a more appropriate name, and Zwicky called them Supernovae (Fig. 5).

What could be the central engine that supplied about 10^{52} erg in a short time? Zwicky and his illustrious colleague Walter Baade, also at CALTECH, were convinced that nuclear fusion, conjectured by Eddington as the source of energy in the stars, could not be the origin of the energy released in a supernova. In one of the most prescient papers published in the entire history of astronomy, Baade and Zwicky proposed in 1934 that the source of energy must be related to the formation of a neutron star. They hypothesized that if a star suddenly collapsed to a radius of the order of 10 kilometres, then the gravitational potential energy released would be of the order of 10^{52} erg! They went on to conjecture that such a collapsed star would consist essentially of neutrons, and they called them *neutron stars*! This was an extremely ingenious idea, coming so soon after Chandrasekhar's discovery in 1930, and the discovery of the neutron in 1932!

In the 1940s, it was pointed out that the position in the sky of the Crab Nebula coincided with the guest star of 1054 AD. It had already been established that the Crab Nebula was expanding very rapidly with a velocity of nearly 1500 km per second. An extrapolation of this expansion *backwards in time* confirmed that the nebula was the expanding debris of the supernova explosion of 1054. Soon after the discovery of the first neutron star in 1968, observations revealed the presence of a neutron star right near the centre of the Crab nebula. This was a spectacular confirmation of the conjecture by Baade and Zwicky that the birth of a neutron star is responsible for the supernova explosion. There was only one missing link. Remember that the original star consisted of an equal number of *protons and electrons*. In the mid 1930s, Lev Landau pointed out that when a star collapsed to a density of the order of 10^{11} g cm^{-3}, electrons will combine with protons to form neutrons. In the process, very weakly interacting particles known as neutrinos would be emitted. Since there are roughly 10^{57} protons in the Sun, when the star collapses to become a neutron star, roughly 10^{57} neutrinos would be emitted as a burst, but all this was mere theoretical conjecture in the mid 1930s, since the neutrinos were discovered only in 1956.

In February 1987, a massive star exploded in the neighbouring galaxy known as the Large Magellanic Cloud. The explosion was first seen as a guest star with the naked eye, and then studied with powerful telescopes. Miraculously, a giant underground neutrino detector in Japan detected a burst of neutrinos *at the same time*! One could trace the direction from which the neutrinos came, and that pointed to the position of the exploding star. That, plus the coincidence in time, confirmed that the burst of neutrinos was associated with the supernova, thus confirming the last missing link in the story!

The investigation of supernovae—both theoretical and observational—was one of the exciting areas of research in the closing decades of the last century. This was important not just for understanding why and how a star explodes. There was another reason. While the lighter elements like *deuterium, helium* and *lithium* were synthesized when the Universe was very young (roughly three minutes after the beginning), the heavier elements are not of cosmological origin. It is widely believed that they are synthesized in stars. To verify these predictions in detail, one will have to study the spectrum of radiation from the ejecta of the explosion. Since the ejecta expanding with a velocity of the order of *several thousands of kilometres per second* will be at a temperature of nearly hundreds of million degrees, the radiation emitted by it will be at X-ray wavelengths. To derive the chemical composition of matter synthesized in the star, one will need to study the *spectrum* of the X-rays emitted by the ejecta (The principle is the same as the one used by Kirchoff and Bunsen to unravel the chemical composition of the Sun. In that case, they used the spectrum obtained by Fraunhofer at visible wavelength). Several sophisticated X-ray telescopes were launched in order to study the X-ray spectrum very accurately. The most impressive amongst them was the giant X-ray observatory CHANDRA (named after the famous astrophysicist *Subrahmanyan Chandrasekhar*) launched by NASA. Thanks to its unprecedented sensitivity,

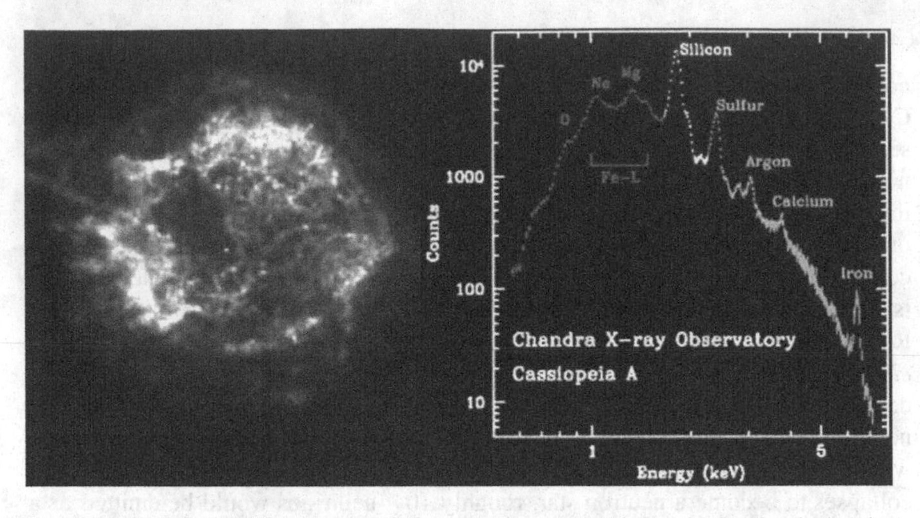

Fig. 6 On the left is an X-ray image of the supernova remnant known as Cassiopeia A, taken with the imaging telescope onboard the Chandra X-ray Observatory launched by NASA. On the right is the spectrum of the X-rays. X-ray emission lines from various atoms present in the ejecta are clearly seen. [Courtesy of the Chandra X-ray Observatory and NASA]

angular resolution, as well as spectral resolution, this observatory has been able to produce spectacular images of supernova debris in the various X-ray emission lines. These images enabled one to analyze both the composition, as well as the spatial distribution of the elements with unprecedented accuracy (Fig. 6).

Neutron Stars

In 1932, the neutron was discovered. As mentioned above, two years later, Walter Baade and Fritz Zwicky at the California Institute of Technology hypothesized stars consisting essentially of neutrons. The idea was that if one compresses a star to a density of 10^{14} g cm^{-3}—the density of the nuclei of atoms we are familiar with—then the repulsive force between the neutrons, practically touching one another, would prevent any further compression, and the star will be stable. It is just like packing a football with ball bearings; when the ball is jam-packed, it will be quite incompressible. This hypothesis led physicists to the conclusion that even stars more massive than 1.4 $M_{\odot}$ could find ultimate peace, not as white dwarfs but as neutron stars. A density of 10^{14} g cm^{-3} implies that such stars would be incredibly small, with a radius of mere 10 kilometres!

So it appeared that all stars will find peace after all, either as white dwarfs or neutron stars, but this turned out to be not true. Chandrasekhar's discovery implied that there must be a maximum mass for neutron stars also. The maximum mass of neutron stars was discovered by Oppenheimer and Volkoff in 1938. We shall turn to the consequence of this discovery in the next section.

Neutron stars were eventually discovered in 1968 by a young student by name Jocelyn Bell at Cambridge University. Today, their population is nearly a thousand. In the intervening years between their prediction in 1934, and their discovery in 1968, many of their properties had been theoretically predicted. All of these have now been verified by actual observations. Their masses are remarkably close to 1.4 $M_{\odot}$, which, you will recall, is the *Chandrasekhar limit* for white dwarfs. They are endowed with incredibly strong magnetic fields of the order of 5×10^{12} gauss. They are very rapid rotators, the fastest among them spinning 640 times a second! Their period of rotation can be measured with remarkable precision. To give an example, the rotation period of the neutron star known as PSR 1937+21 has been measured to be P = 0.001557806472448830(3) seconds. Indeed, the accuracy with which the rotation period can be determined is limited only by the precision to which the unit of time itself is presently defined!

You will recall that a magnetized rotating sphere is a dynamo. A dynamo, such as the one you may have in your cycle, generates a few volts—enough to light a bulb. Given the strong magnetic field and rapid rotation, neutron stars function as incredibly powerful dynamos; the voltage drop between the poles and the equator can be as large as 10^{16} volts! Not surprisingly, many exotic phenomena can occur near the surface of neutron stars. As a result, rapidly rotating neutron stars emit electromagnetic radiation over a broad spectrum—*radio, infrared, visible, ultraviolet, X-rays and gamma rays*—due to a variety of processes.

Many neutron stars are in binary systems with a gaseous star like the Sun as a companion. When matter from the gaseous companion is sucked by the neutron star's strong gravity and falls on its surface, the neutron star will turn into a powerful X-ray source. The reason for this is quite simple. Since the radius of a neutron star is a mere ten kilometres, the surface gravity is incredibly strong. The gravitational potential energy of a mass m at the surface of the star is approximately equal to 10 percent of its rest *mass energy* mc^2! This means that if we drop a mass m onto a neutron star, the potential energy released is a staggering 0.1 mc^2. This energy comes out mainly as X-rays. This is why neutron stars in binary systems, accreting from their gaseous companion, are powerful X-ray sources. Many hundreds of such X-ray-emitting neutron stars in binary systems are now known, a large number of them in external galaxies.

Black Holes

Albert Einstein published his *General Theory of Relativity* in 1915. Within a year after that Karl Schwarzschild, the great German physicist and astronomer, found an exact solution to Einstein's equations describing the geometry of spacetime inside and outside of a spherical nonrotating star (Einstein had thought that it was highly unlikely that anyone would be able to find an exact solution to his theory!). This was an incredible intellectual feat by Schwarzschild. One of the remarkable features of this solution was the following. When a star shrinks to a critical radius given by $R_s = 2\ GM/c^2$, where M is the mass of the star and c is the velocity of light, no signal can escape from the surface of the star. The wavelength of the radiation is stretched to infinity and the frequency goes to zero; in other words, time comes to a standstill. Hence the star will appear to be black. These are the famous *Black Holes* of General Relativity. This result was so extraordinary that most people did not believe it. Certainly not Einstein! As late as 1938, Einstein published a paper arguing that such objects could not exist in reality. Einstein's paper was mathematically exact, but his physical reasoning was flawed.

As mentioned in the previous section, Oppenheimer and his student Volkoff had established the existence of a maximum mass for neutron stars, the analogue of the Chandrasekhar limiting mass for white dwarfs. The existence of a maximum mass for neutron stars raised the awkward question once again *'what is the fate of stars more massive than the maximum mass of white dwarfs and neutron stars?'* Chandrasekhar had already provided a definitive answer to this question way back in 1932. He had established in a mathematically rigorous fashion that sufficiently massive stars are doomed. They cannot be stabilized by quantum mechanical pressure, no matter how high the density becomes. Such stars have no option but to collapse till they become singularities; an appeal cannot be made to quantum physics to save them. But, as we shall discuss in the next volume, this conclusion was rejected by Eddington, the High Priest of astronomy. In 1939, Oppenheimer and his student Snyder re-examined Schwarzschild's result and came to the same conclusion. But no one took all this seriously. The general opinion was that this extraordinary result must be an artefact of certain simplifying assumptions made along the way.

In 1963, all this changed when astronomers discovered objects which have come to be known as Quasars. The remarkable thing about these objects was that the amount of energy they radiate per unit time was staggering. Quasars radiate approximately 10^{46} erg/sec, which is 10^{13} times the energy the Sun radiates per second. To put it differently, *the energy radiated by quasars in one second is equal to what the Sun would radiate in a million years*! This implied that quasars must be at least a billion times more massive than the Sun (We shall discuss the underlying reason in Chap. 3, 'Eddington's theory of the stars', in this volume). Further observations led one to the conclusion that this enormous mass must be concentrated into such a small radius that it must be a black hole! Soon a new paradigm emerged. *The central engines of the countless number of quasars that populate the Universe must be supermassive black holes.*

The first concrete evidence of *stellar mass black* holes came in 1973. A powerful binary X-ray source was discovered in the constellation Cygnus, with a compact object and a gaseous star going around a common centre of mass. This was christened as Cygnus X-1. Unlike in so many other cases of binary X-ray sources, the compact object in this case could not be a neutron star. Using Kepler's laws, one could estimate the mass of the compact object, and it came out to be about six solar mass—much larger than the maximum mass a neutron star could have. After the pioneering effort in 1938 by Oppenheimer and Volkoff to estimate the maximum mass of neutron stars, the question was revisited by a number of researchers. By the mid-1970s there was a consensus that the maximum mass of neutron stars must be around $2M_{\odot}$; in any case, there were compelling reasons to assert that it could not be more than about $3.2M_{\odot}$. Therefore, the compact member of Cygnus X-1 could not be a neutron star, and it goes without saying that it cannot be a white dwarf. *It had to be a black hole*. Admittedly, this was circumstantial evidence. Nevertheless, it was a compelling argument.

The discovery of quasars and Cygnus X-1 triggered a revival of interest in General Theory of Relativity. A number of exceedingly bright young people were attracted to the study of black holes. Roger Penrose and Stephen Hawking were among them. During this period of intense research, a number of *exact theorems* were proved concerning the properties of black holes. The culmination of all this was the spectacular discovery made by Stephen Hawking in 1974. He discovered that particles and radiation could get out of black holes, after all! According to him, a black hole will emit particles and radiation, just as a black body emits radiation. This was not a violation of General Relativity. *Particles and radiation tunnel out of black holes due to quantum mechanical processes, just as alpha particles escape from the atomic nuclei*. This discovery by Hawking was universally hailed.

These theoretical efforts, in turn, led to more intense search for black hole candidates. The list of stellar mass black holes in binary systems has grown to a substantial number. Highly sensitive X-ray telescopes, such as the Chandra X-ray Observatory, have been able to detect such sources in other galaxies, as well. As the last century ended, there is mounting evidence for supermassive black holes at the centres of most galaxies. Although the evidence is somewhat circumstantial at the moment, most astronomers believe that this is the most conservative hypothesis to explain the observational results. From an observational point of view, the evidence for a black hole can only be considered as conclusive if one is able to detect effects which are *manifestly General Relativistic in nature*, for after all black holes are predictions of General Relativity, but one may have to wait for some more time for this.

Between the Stars: The Interstellar Medium

When you look at the night sky, the space between the stars appears empty, but this is not the case. *In a galaxy like ours, there is as much matter between the stars as locked up in the stars*! This is known as the interstellar medium, and the discovery of its structure and constituents was one of the major achievements of the last century. Interestingly, one had to wait for the advent of the quantum theory of matter, as well as new windows to the Universe opening up, to unravel the mysteries of this medium.

The interstellar medium is like a *raisin pudding* (Fig. 7). There is a diffuse medium, consisting mainly of atomic hydrogen. The density of this tenuous gas is

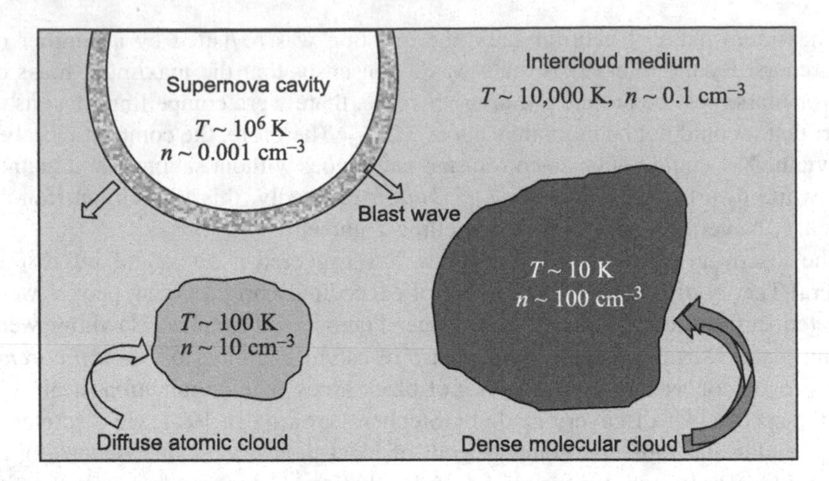

Fig. 7 The raisin-pudding model of the interstellar medium (ISM). The ISM consists of diffuse hydrogen gas. Embedded in this gas are gigantic atomic and molecular clouds. The expanding blast waves from Supernovae excavate gigantic cavities in the ISM. These cavities are filled with gas at millions of degrees

approximately 0.1 atoms per cm^3 and its temperature is about 10,000 kelvin. This density may seem absurdly small to you. If you want to have some fun, compare this with the number of molecules per cubic centimetre of water, air, and the best manmade vacuum, respectively. Yes, the density of the interstellar medium is unimaginably small. But given the enormous volume of the galaxy, it amounts to a great deal of gas!

Embedded in this diffuse medium are giant clouds of gas (these are the raisins and dried fruits in the pudding). There are two kinds of clouds. The more diffuse clouds consist essentially of atomic hydrogen. The denser clouds are almost entirely molecular in nature, with hydrogen molecule being the most abundant. These dense clouds also have a lot of dust particles. Consequently, these molecular clouds are opaque to visible light. And they are really giant clouds. They are many tens of light years in size, and contain as much as 10 million solar mass of molecular gas and dust.

The interstellar medium is not a peaceful place. Indeed, it is a very violent arena! What destroys its serenity are the supernova explosions of stars, which happen once every 40 to 50 years. The ejecta of the exploding star initially expand at incredible speeds, which could be as great as 15,000 *kilometres per second*! This creates a powerful shock wave which propagates in the interstellar medium. The interstellar matter that is swept up by the expanding spherical shock wave is heated to a temperature of several million degrees. Such a hot supernova bubble will continue to expand till its motion is eventually arrested by the external pressure exerted by the undisturbed interstellar medium. Such bubbles, filled with highly tenuous million degree gas, will last for many millions of years. Indeed, these super-hot bubbles may connect up with one another to form tunnels in the interstellar medium. Thus, our nice raisin pudding will look more like Swiss cheese! As the expanding hot super bubbles engulf the diffuse clouds, they will slowly evaporate them. When the expanding supernova blast waves impact on the giant molecular clouds, they will trigger the formation of young stars. As the hot gas in the bubble eats away the molecular gas, the newly formed stars will be slowly revealed. Thus, the interstellar medium is a very dynamic place, witnessing the birth and death of stars, as well as the birth and destruction of interstellar clouds.

The Birth of New Stars

If stars are dying so frequently, is the stellar population being constantly replenished by newly born stars? I am sure that this question seems perfectly reasonable to you, and yet, the *birth of new stars* was not considered a meaningful question till the 1950s. The first evidence of very young stars, barely a *few million years* old, came with the discovery of certain runaway massive stars moving with great speeds. This news was received with great consternation. The Galaxy is many billions of years old. How could there be stars with ages of the order of a few millions of years? This clearly pointed to the fact that stars are *still being born*. The extrapolation of the velocity vectors of these *runaway stars* backwards in time, led one to their birth place! Another question that was considered frivolous not so long ago had suddenly acquired meaning. Indeed, a detailed understanding of the birth of stars is one of the very active areas of research in astronomy today.

Soon there was ample evidence to show that new stars are born when giant molecular clouds of gas collapse. As mentioned above, these giant molecular clouds are opaque to visible radiation. To probe their interior, one will have to use either millimetre waves or infrared radiation. Since their wavelength is larger than the size of the dust particles, these waves can escape from these clouds rather easily (Fig. 8).

Although millimetre wave astronomy came of age in the 1970s, it proved to be useful mainly to study the radiation emitted by the molecular material out of which stars are born. To study the emission from the newly forming star itself, one had to wait for the advent of infrared astronomy.

Although we are now able to see stars being born, many of them with gaseous discs around them, from a theoretical point of view there is much to be understood. For example, one still does not understand in detail how the collapsing cloud manages to shed its angular momentum. You know from the example of the ice skater that she spins up dramatically as she pulls in her hands. This is a consequence of the conservation of angular momentum. In a similar fashion, a slowly rotating cloud will spin up as its radius decreases. At some stage the rotation becomes a hindrance to the collapse itself due to the centrifugal force. Therefore, the collapsing cloud has to find an effective mechanism to shed its

Fig. 8 These gigantic pillars of molecular gas and dust are part of a nearby star-forming region about 6,500 light years away. Buried inside such opaque clouds are newly forming stars. To view them, one has to look in the infrared. Courtesy of NASA, ESA, STScI, J. Hester and P. Scowen (Arizona State University)

angular momentum. Otherwise the collapse cannot proceed all the way to form stars. The fact that stars are forming tells us that nature is ingenious enough to find ways. But we do not yet know in detail what the mechanism is (Fig. 9).

Even if we set aside these difficulties, there are other fundamental questions to be answered. Stars have a range of masses, with the majority of them having masses between half the mass of the Sun to about ten times its mass. Within this range, the number of stars decreases with increasing mass, as a power law. A major fraction of the more massive stars are in binary systems. Young stars of the same age are seen to belong to clusters. Therefore, we need to know what determines the spectrum of the masses of the stars formed by the collapse of a giant cloud of gas, why some stars belong to multiple systems and not others, and other such details.

Was There a Beginning to Spacetime?

We mentioned earlier that George Gamow was one of the very few persons who realized the importance of Hubble's discovery of the recession of the clusters of galaxies. Extrapolating backwards, he estimated that when the Universe was about one second old, its temperature would have been of the order of 15 billion degrees.

Fig. 9 NASA images of Trapezium cluster in Orion Nebula taken with the Hubble Space Telescope. The visible image is on the left and infrared image on the right. The visible image shows the brilliant massive stars which are part of the Trapezium cluster; they are so bright that they saturate the photograph. On the right is an infrared image of the same region. The numerous tiny bright spots are newly formed stars buried deep in the opaque cloud. [These images were created for NASA by Space Telescope Science Institute and for ESA by the Hubble European Space Agency Information Centre. Courtesy of NASA, ESA and STScI]

Gamow realized that a hot expanding Universe was the ideal laboratory for synthesizing the light elements such as deuterium, helium, lithium etc. He also appreciated that since matter and radiation would have been in true thermodynamic equilibrium, the primeval radiation would have had the characteristics of black body radiation; its spectrum would be as predicted by Planck's Law. By 1948, Gamow and his student Ralph Alpher had worked out an elaborate theory of nucleosynthesis. This was followed by a more careful analysis by Ralph Alpher and Robert Herman (1948). An important prediction of these papers was the following. As the Universe continued to expand, the primeval radiation would have cooled due to adiabatic losses, but would have retained its black body nature. They predicted that the present Universe must, therefore, be filled with the relic of this primeval fireball, and that its temperature would be roughly a few degrees kelvin.

In 1964, Arno Penzias and Robert Wilson, working at the Bell Labs in Homdel, NJ, USA, accidentally discovered this relic radiation. But they measured the intensity of the all pervading radiation at only one frequency. Since this radiation was present everywhere, and since its intensity was the same in all directions, it was suggestive of black body radiation. Assuming it to be true black body radiation, they estimated a temperature of 3 kelvin. In 1978, Penzias and Wilson were jointly awarded the Nobel Prize for Physics for this monumental discovery.

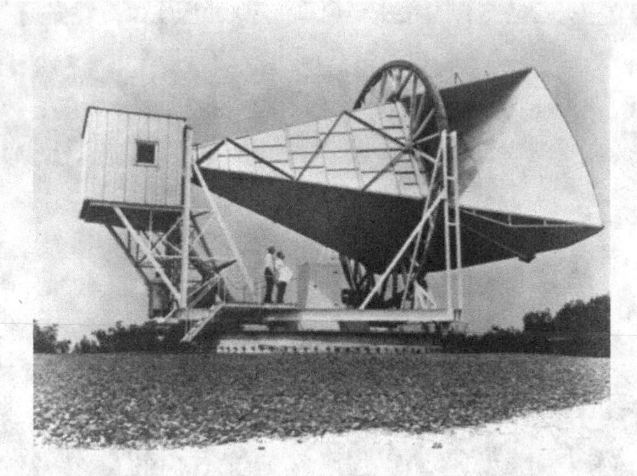

Fig. 10 Penzias and Wilson standing next to the famous 'Horn' radio telescope at the Homdel Bell Labs, with which they made the historic discovery of the Cosmic Background Radiation

Although most people were convinced that Gamow had got the story right, it was still extremely important to establish that the cosmic background radiation was really black body radiation, and to do that, one had to measure its spectrum over a wide range of frequencies and demonstrate that it obeyed Planck's Law. But this could not be achieved from ground-based observations because of atmospheric absorption in crucial wavelength bands. Some observations were attempted from high altitude balloons, but they were not very accurate. Towards the end of the last century a satellite called COBE—Cosmic Background Explorer—was launched to clinch this issue (Fig. 10). And it did! Precise measurements showed that the cosmic microwave background radiation has a perfect *planckian spectrum*. The temperature of this radiation was measured to be 2.725 kelvin. The background radiation was definitely of cosmological origin. But did it prove that the Universe had a beginning? Not quite.

Most physicists were still very sceptical. The discovery of this radiation did not prove that the Universe had a beginning. It is important to understand the reason for this scepticism. If one extrapolated backwards the recession of the galaxies, one will naturally encounter a smaller, denser and hotter Universe. But how far is this extrapolation believable? And what is the framework for such an extrapolation? We shall not digress now to discuss this, but return to it in the last volume of this series. I request you to accept the following statement for the moment. Einstein's General Theory of Relativity admits the possibility of an expanding or contracting Universe. The equations of the General Theory of Relativity tell us how the temperature and the density of the Universe will change with the characteristic size of the Universe. Using these equations, one can extrapolate backwards, and infer the temperature and density at an earlier epoch. For example, the discovery of the Cosmic background radiation with a temperature of 3 kelvin

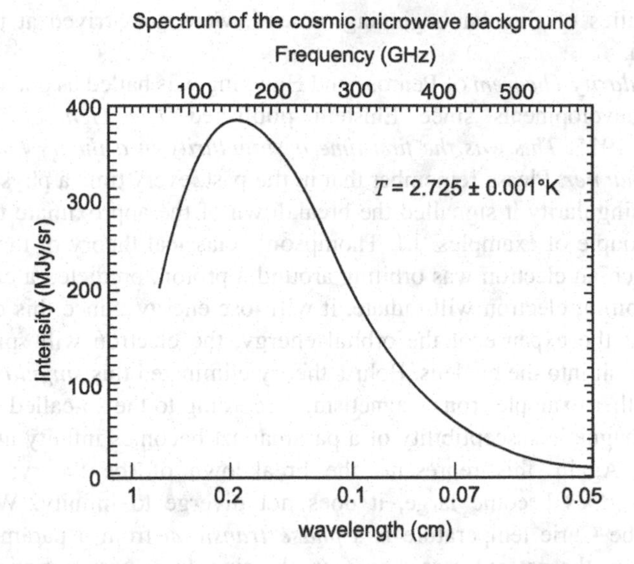

Fig. 11 The cosmic microwave background spectrum measured by the FIRAS instrument on the COBE satellite is the most-precisely measured black body spectrum in nature. The data points and error bars on this graph are smaller than the thickness of the theoretical curve! [Courtesy of NASA and the COBE science team which produced this incredible spectrum]

tells us that when the characteristic size of the Universe was a *thousand times less than* what it is now, the temperature of the Universe was 3,000 kelvin; the black body nature of the radiation tells us that matter and radiation were in equilibrium at this temperature. The age of the Universe at that epoch would have been roughly 300,000 years. At such a temperature, and the density that would have obtained then, there is no reason to doubt the laws of physics as we know them. To synthesize the elements, however, Gamow had to extrapolate the equations of general relativity much further back in time. The synthesis of elements, according to him, occurred when the Universe was merely 3 minutes old, and the temperature was roughly a *billion degrees* (Fig. 11).

Indeed, a further extrapolation leads to zero radius and *infinite density* at a finite time in the past; in other words a singularity! Can we then take it that according to General Relativity there was a beginning to the Universe? No, because the above conclusion was based on an *extrapolation*. Therefore, whether the Universe had a beginning continued to remain a speculative question since there was no way to pose this rigorously within the premise of General Theory of Relativity.

This changed dramatically in 1965, when Roger Penrose and Stephen Hawking proved an exact theorem, which has come to be known as the Singularity Theorem. They showed in a mathematically exact manner that if some reasonable conditions are assumed, Einstein's *Theory of General Relativity* predicts that the *Universe could have either begun in a singularity, or end in a singularity, or both began and will end in a singularity.* It is important to stress that this was an exact theorem.

The possibilities of singular beginning, or end, were not arrived at through an extrapolation.

The *Singularity Theorem* of Penrose and Hawking was hailed as one of the most significant developments since Einstein published *The General Theory of Relativity* in 1915. *This was the first time a singularity in a theory had emerged in an exact manner.* Please remember that in the past, every time a physical theory predicted a singularity it signalled the breakdown of the approximate theory. Let us recall a couple of examples. J.J. Thompson's classical theory of the hydrogen atom, in which an electron was orbiting around a proton, predicted a catastrophe. Since the orbiting electron will radiate, it will lose energy. Since this energy can only come at the expense of the orbital energy, the electron will spiral in and eventually crash into the nucleus. Bohr's theory eliminated this *singularity*. Let us consider another example from magnetism. According to the so-called mean-field theory, the magnetic susceptibility of a paramagnet becomes infinity at the Curie temperature. Again, this represents the breakdown of the theory; while the susceptibility does become large, it does not diverge to infinity. What really happens at the Curie temperature is a *phase transition* from a paramagnet to a ferromagnet. In the present case, however, the singularity was not an artefact of any approximation made to *The General Theory of Relativity*. It is an exact result. Its acceptance or rejection can only be made on philosophical grounds!

At any rate, the theorem of Penrose and Hawking made the grandest of all questions, namely, *Was there a beginning to spacetime?* a meaningful scientific question, one which could now be posed rigorously within the premise of General Relativity.

Astronomy at the Dawn of the Millennium

As we enter the new millennium, we are confronted by many new outstanding questions. The exciting thing is that despite the enormous progress made in the last century, there are more unanswered questions now than ever before. Interestingly, these questions cover the entire range of objects in the Universe ranging from the smallest to the largest: planets, stars, galaxies, clusters of galaxies, and the Universe at large. Before concluding this overview, let me briefly mention a few examples, just to whet your appetite.

Extrasolar Planets

One of the most exciting areas in astronomy today is the search for planets around other stars. Finding extra solar planets is important to understand the formation of the planets in the solar system and, indeed, the solar system itself. Ideally, one would like to find planetary systems around *young stars*. Curiously, the first exoplanet was discovered in 1992 orbiting a neutron star—the remnant of a dead star! Since the birth of a neutron star is accompanied by a supernova explosion, it is not at all clear how this planetary system survived the violent explosion!

The first definitive detection of an *extrasolar* planet was in 1995 orbiting a star known as 51 Pegasi. As of January 2010, forty-five multiple-planet systems are known and the number of extrasolar planets has grown to 429! The vast majority of exoplanets detected so far are gas giants (presumably like Jupiter). As of January 2010, all but twenty-five of them have more than ten times the mass of Earth. Many are considerably more massive than Jupiter, the most massive planet in the Solar System. It is most likely that this is due an observational selection effect: massive planets are easier to find. Planets with a mass smaller than about 0.75 times the mass of Jupiter orbit very close to the parent star. These findings are presumably telling us something about the migration of planets. Another exciting recent discovery is that invariably the host stars are quite metal rich. The frequency of planets increases with the iron content of the stars. It is not clear why this is so.

There is another reason why this search for extrasolar planets, particularly *Earth-like* planets, is being pursued so vigorously. Mankind has always been intrigued by the question, *Is there any life out there?* There are several major programmes to detect 'messages' from alien life. There have also been attempts to transmit radio signals, coded to contain information about us and the world we live in. The hope is that these signals may some day be detected by extraterrestrial life. A more meaningful approach might be to look for small planets like the earth, orbiting the parent stars at the right sort of distance where conditions favourable for life might exist. At the moment, planets that have been found in the so-called *habitable zone*, where Earth-like conditions may prevail, are gas giants like Jupiter. Life cannot exist in these gas giants! But if some of these planets have moons, life could perhaps exist in one of the moons.

In March 2009, NASA launched a spacecraft named **KEPLER** (named after Johannes Kepler). This spacecraft has the sensitivity to detect both small and large planets. By January 2010, the Kepler science team had announced the discovery of five new planets, ranging in size from four times the size of the Earth to larger than Jupiter. They have orbits ranging from 3.3 to 4.9 days. Estimated temperatures of the planets range from 2,200 to 3,000 degrees, hotter than molten lava and much too hot for life as we know it. In the next couple of years, this mission is bound to find many more planets. Hopefully, some of them will be Earth-like planets orbiting their parent stars in the habitable zone. These are early days still. With many more missions planned we are in for many surprises.

The Origin of Galaxies

This question is at the base of the present revolution in astronomy. And yet, just a few decades ago, this would have been considered as a meaningless question. The possibility that galaxies always existed must surely be rejected if one accepts that the Universe had a very hot beginning. At the temperatures that existed in the early Universe the only sensible thing to expect is a primordial soup of a variety of elementary particles and radiation, in true thermodynamic equilibrium.

Gravitationally bound objects like galaxies must have formed from the primeval homogeneous matter as the Universe expanded and cooled. Although this was pointed out by Lemaître soon after Hubble's discovery of the recession of the clusters of galaxies, one began to take this seriously only after the discovery of the comic background radiation.

If galaxies formed at some epoch as the Universe expanded and cooled, then there must have been precursors in the form of density fluctuations in the early Universe. Analogy with the behaviour of a gas as it is cooled towards the liquefaction temperature might make this plausible. A gas in thermodynamic equilibrium will be of uniform density. As the gas is cooled towards the transition temperature, density fluctuations appear, and they grow in amplitude. Soon the regions of enhanced density can be identified with *droplets of water*. There will be a spectrum of droplet sizes, and they will appear and disappear. As the phase transition is approached, the fluctuations will slow down and the characteristic size of the droplets will increase. One may expect the formation of galaxies to mimic this. While this analogy might be used to guide intuition, one must bear in mind that in the present case the physics is very different. Coming back to how galaxies might have formed, the idea was that density fluctuations might have spontaneously appeared as the Universe expanded. Indeed, there were predictions of manifestly quantum fluctuations when the Universe was about 10^{-35} seconds old. These fluctuations could grow, initially linearly and then nonlinearly. Galaxies might have been born out of these primordial density fluctuations.

If one accepts this general premise then one must ask if these density fluctuations might have left *imprints* in the Universe. Unfortunately, the Universe was opaque till the age of about 300,000 years. Therefore, that is the earliest epoch we can see. Indeed, when we look at the cosmic background radiation, we are looking at the epoch when the Universe was roughly 300,000 years old. Is there any evidence of the primordial density fluctuations when we look at this epoch? If the density fluctuations grew adiabatically, then one would expect regions of density enhancement to be hotter, and regions of density deficit to be cooler than the average. This leads to a clear prediction if one accepts this broad scenario: *since matter and radiation were in equilibrium, the temperature of the cosmic background radiation must show fluctuations over the sky*. The theoretical scenarios predicted a fluctuation in the temperature of the radiation at the level of tens of micro kelvin. After sustained effort, the COBE satellite finally found the temperature fluctuations, and these were at the level of a few tens of *micro kelvin*. A few years later, another spacecraft called WMAP was launched to detect finer details in the anisotropy. This is shown in Fig. 12.

While the discovery of the anisotropy in the temperature of the cosmic background radiation lends credibility to the idea that galaxies could have grown from the primordial density fluctuations, one is a long way from understanding the process of galaxy formation in detail. Matters have been made more difficult by the

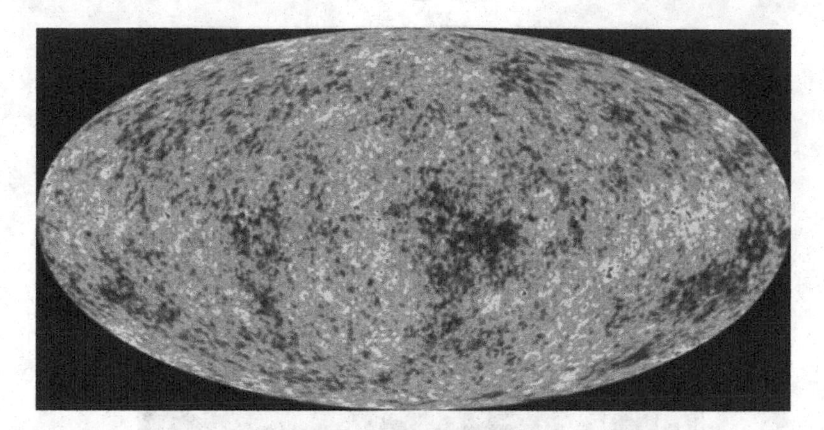

Fig. 12 The Cosmic Microwave Background temperature fluctuations from the 5 year Wilkinson Microwave Anisotropy Probe data seen over the full sky. The average temperature is 2.725 Kelvin, and the light grey and black patches represent the tiny temperature fluctuations, as in a weather map. The hotter and colder regions differ in temperature by about 0.0002 degrees. [Courtesy of NASA/WMAP Science Team]

recent discoveries of more and more distant galaxies. The most distant galaxy discovered so far is at a redshift of about 7. The characteristic size of the Universe at that epoch would have been one-seventh of the present size, and the age of the Universe not more than about 750 million years. For galaxies to form and become tightly bound, a deep gravitational potential well has to form and gas has to fall into it. The embarrassing thing is the following. When the first galaxies formed, the Universe was less than about a billion years old. That is not a long enough time for matter to free fall to form a compact galaxy. So the early galaxies might not have been as compact as the present galaxies which are typically about 100,000 light years across. The early galaxies may not have had flattened and rotationally stabilized disks like our Galaxy. The interstellar medium in the sense we see in nearby galaxies, may not have formed. A related question is '*when did the first stars form*?' The spectrum of the most distant quasars suggests that some stars should have formed even before galaxies formed (Fig. 13).

In the nearby Universe, galaxies have several distinct shapes. Some are flattened spirals, some with bars at the centre; some are spheroidal and so on. The observed shapes were classified into a sequence nearly a hundred years ago. And yet, when we look deep into the Universe with the Hubble Space Telescope, we see galaxies with highly irregular shapes. Collisions and mergers of galaxies seem to be a common phenomenon. So both the origin of galaxies and their evolution remain very challenging questions.

Fig. 13 This pair of galaxies, known as the *Mice* for their tails of stars and gas, have collided and will eventually merge into a single galaxy. Streams of material have been tugged out of the galaxies by the force of gravity, triggering new star formation. [Credit: NASA, H. Ford (JHU), G. Illingworth (UCSC/LO), M.Clampin (STScI), G. Hartig (STScI), the ACS Science Team, and ESA]

Dark Matter in the Universe

The first clue that not all of the mass in the Universe is 'luminous' came nearly seventy years ago from a study of the clusters of galaxies. It is believed that such clusters are gravitationally bound. The individual galaxies in a cluster are moving around the centre of the self-consistent gravitational potential. In other words, every galaxy is moving in the potential well due to all other galaxies. The velocity of an individual galaxy, and therefore its kinetic energy, can be measured spectroscopically using the Doppler shift of known spectral lines. By adding these up, one can estimate the total kinetic energy of the galaxies. Given this estimate, one can deduce what the gravitational potential energy of the cluster of galaxies *ought to be* for the system to be bound. This can be done by invoking the well-known *virial theorem* which says that the *kinetic energy must be equal to one half the gravitational potential energy*. Given the size of the system, the gravitational potential energy is determined by the total mass enclosed (P.E. $\sim GM^2/R$).

It is an extraordinary fact that in all the clusters studied so far the total amount of luminous mass is only ~ 10 % of the mass required to gravitationally bind the clusters. This was pointed out more than 70 years ago by Fritz Zwicky. One had hoped that the ticklish problem will go away, but it has not! This is the famous dark matter problem. The mass of the cluster is really ten times more than what we can deduce from the radiation they emit.

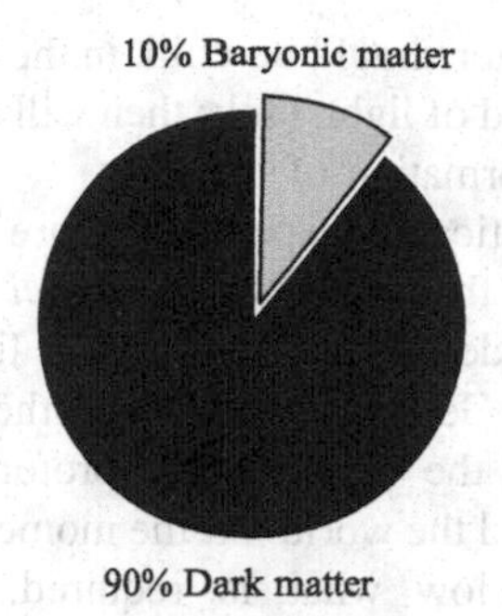

Fig. 14 One of the great surprises of contemporary astronomy is that only about 10 % of matter in the Universe is baryonic matter that we are familiar with. The remaining roughly 90 % is of unknown nature. All we can say is that the constituents of the dark matter must be extremely weakly interacting massive particles. But this unseen matter which pervades the Universe makes its presence felt through its gravitational interaction with the baryonic matter

There is a similar problem in individual galaxies themselves. It is now well established that the major fraction of the gravitational mass of galaxies is *dark*. The unseen mass in galaxies is again of the order of 90 % of the total mass. The observational evidence that led astronomers to this inescapable conclusion is different from the one mentioned above. But we shall not pause to discuss it here.

The bottom line is the following. *Based on a variety of observations, astronomers have established beyond reasonable doubt that galaxies must have a huge and extended halo of dark matter. This halo of dark matter accounts for nearly 90 % of the total mass of the galaxy.*

One of the most outstanding questions of contemporary physics is the nature of this unseen matter which makes its presence felt only through its gravitational influence, and which pervades the Universe. While some of the dark matter may be accounted for by invoking planets, black holes and the like, most of it must be *nonbaryonic* in nature. Baryonic matter is matter that we are familiar with—made of neutrons and protons. If more than a few percent of the dark matter was baryonic, then there will be a serious contradiction with the observed abundance of light elements such as deuterium and the lighter isotope of helium. These elements are believed to have been synthesized when the Universe was just a few minutes old, and the observed abundance seriously constrains the amount of baryonic matter present in the Universe at that time (Fig. 14).

Although the constituents of this nonbaryonic matter is not known, some of the Grand Unified Theories of Physics predict a class of Weakly Interacting Massive Particles (WIMPS) which might have been produced in large numbers in the early Universe, and which dominate the present day mass content of the Universe. An example of such a particle predicted by theory, but not yet detected, is the *Neutralino*. This is supposed to have a mass about 100 times the mass of the proton. But, like the neutrino, it will interact extremely weakly with normal matter. The important thing is that such particles will satisfy the requirement of the astronomers that the dark matter should be cold, in the sense that its velocity must be small compared to the speed of light. Only then will such particles aggregate (or condense) and assist in the formation of galaxies.

A variety of very sophisticated experiments are underway to detect such particles. One of them is in the *Gran Sasso Tunnel* in the Alps Mountains in Europe. This has already yielded a significant lower limit in that the mass of the WIMP must be more than 50 GeV. In other words, the particle one is looking for must be at least fifty times the mass of the proton. There are several other experiments in progress around the world. At the moment the sensitivity of most of these experiments is far below what is required. But given the profound significance of the detection of these particles—both for fundamental physics and astronomy—new technological initiatives are being explored to greatly improve the sensitivity.

The Dark Energy and the Accelerating Universe

Two questions concerning our evolving Universe were at the centre stage during the last century:

1. Will the Universe expand for ever, or will it start contracting at some stage due to its own gravity?
2. What is the geometry of the Universe?

According to the *General Theory of Relativity*, these two questions are related. Why is the Universe expanding in the first place? Frankly, we do not know. All one can say is that Einstein's *General Theory of Relativity,* which provides a framework to describe the Universe, admits this possibility. And what will be the eventual fate of the Universe? This is a question that concerns the geometry of the Universe, and General Relativity has a precise answer for this. According to General Relativity, *the curvature of space can be zero or positive or negative*. Which of these three possibilities describes our Universe depends on how much matter there is in the Universe. If the density is greater than a certain critical density (whose value is roughly the mass of 5 protons per cubic metre), then the gravity of the Universe will eventually pull back the matter, and the Universe will be finite. The curvature of space would be positive. If, on the other hand, the density is less than the critical value then the curvature of space would be negative. *If the density of the Universe is precisely equal to the critical density then the curvature of space would be zero, and the geometry of space would be Euclidean.* This is like a stone being thrown up with a velocity precisely equal to the escape velocity. The stone will escape to infinity, but it will have zero velocity at infinity.

Well, which of these three possibilities describes our Universe? The only way to settle this is by actually measuring the density. Astronomers have spent several decades trying to measure the density of our Universe. The principle is simple enough. Measure the mass inside a sufficiently large volume, and divide by the

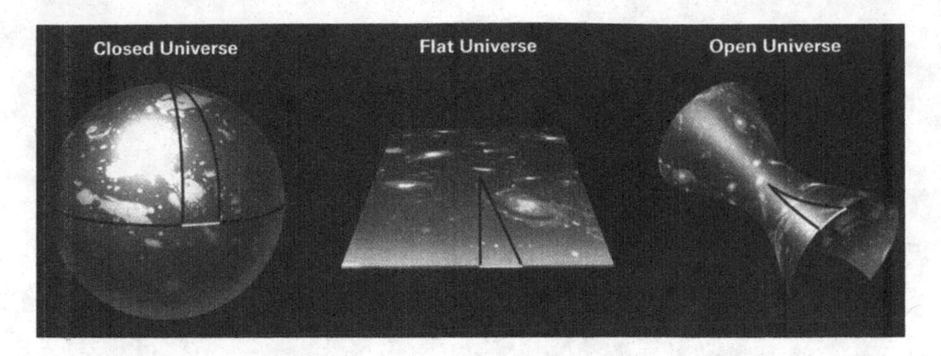

Fig. 15 Three possible geometries of the Universe. Whether the curvature of space will be positive (*left*), zero (*centre*) or negative (*right*) will depend upon whether the density of the Universe is greater than, equal to or less than the critical density

volume! The result of this very elaborate exercise is the following. *The density of the Universe is only about 25 % of the critical density*. This implies that the Universe will not contract at some stage due to its own gravity. The Universe will go on expanding for ever. According to General Relativity, the curvature of space must be negative (Fig. 15).

But there was shocking news from another set of astronomers. Earlier we referred to the measurement of the anisotropy in the microwave background radiation. The first successful measurement of this by the COBE satellite did not have sufficient angular resolution to go beyond discovering the anisotropy. This was followed up by balloon-borne telescopes with extremely sophisticated receivers. These measurements had sufficient angular resolution to draw more quantitative conclusions. This was followed by the launch of a dedicated satellite called WMAP. A detailed analysis of the anisotropy at various angular scales has led one to the amazing conclusion that the curvature of space is zero. To put it differently, the geometry of the Universe is Euclidean. We shall return to this exciting story in the last volume of this series. For now, we shall merely state the following. To measure the geometry of space what we need is a standard rod at a standard distance. The *angle* subtended by this standard rod will tell us what the geometry of the intervening space is. The point is that the anisotropy of the microwave background provides us with a *standard rod* at a precisely known distance. And the angle subtended by this standard rod, at a known distance, was precisely equal to what Euclidean geometry would have predicted.

What is absolutely amazing is that this result was predicted by theoretical physicists nearly twenty years before the observations!

But we now have a very strange dilemma. A direct measurement of the matter density of the Universe told us that the density of the Universe is roughly 25 % of the critical density. And yet, a Euclidean geometry—again, a direct measurement—requires that the density of the Universe must be *precisely equal to the critical density*. How can we reconcile these two statements? We can only do so by

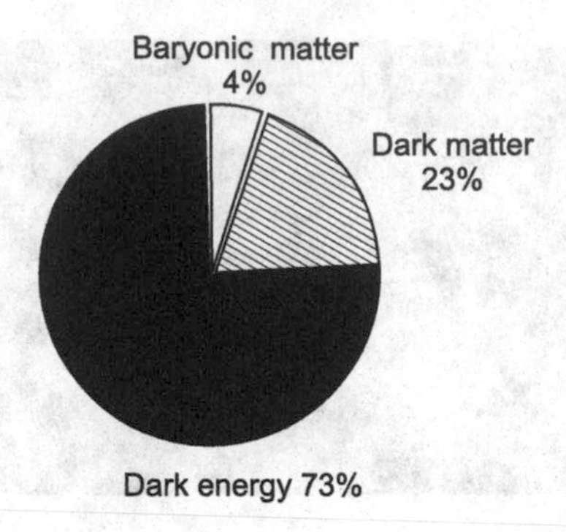

Fig. 16 Even more intriguing than the nature of the dark matter is the mystery of the dark energy. It now appears reasonably certain that roughly 73 % of the energy density of the Universe may not be associated with baryonic or dark matter. This inference is forced upon us by the spectacular discovery that the geometry of the Universe is Euclidean. Since this dark energy is not associated with matter, it can only be associated with vacuum or empty space. The remarkable thing is that although the energy density of vacuum is positive, the pressure exerted by it is negative! It is this negative pressure of the dark energy that is responsible for the observed acceleration of the Universe

attributing *75 % of the energy density of the Universe to empty space or vacuum* (Fig. 16).

This may seem like an innocent way to resolve a major paradox. But it has far reaching implications. Vacuum (or empty space, if you like) is quite an uninteresting arena in classical physics. In quantum physics, however, there is rich physics associated with vacuum. Some of it has measurable consequences in the material world of the atoms. One of the curious things about vacuum is that while its energy density is positive—like in the case of matter—it has negative pressure. This means that the energy density due to the vacuum acts like *repulsive gravity*. This cosmic repulsion due to the vacuum would not have any significant dynamical consequence for the Universe at large if its contribution to the energy density budget had been small. But this is not the case. As we saw, vacuum contributes roughly 75 percent of the energy density in the present epoch. This, in turn, means that the cosmic repulsion dominates over the cosmic attraction due to the gravitation force acting on matter. We are thus led to the extraordinary conclusion that the expansion of the Universe, instead of slowing down due to gravity, should be accelerating. In the analogy of the stone being thrown up from the earth, the stone is not only escaping to infinity, but its velocity, instead of decreasing, is increasing! The Universe is getting curiouser and curiouser.

Yes, our Universe is indeed accelerating! Recently, observation of distant supernovae has shown that the expansion of the Universe is, in fact, speeding up. We shall not digress to discuss this exciting discovery.

This is an extraordinary state of affairs. It was hard enough to reconcile to the fact that matter, of which we are made of, is only a small fraction of the mass in the Universe. Now it is turning out that the Universe is dominated by a strange form of energy which, contrary to everything we know, gives rise to a cosmic repulsion. This mysterious energy has been termed the dark energy! As of now, none of the physical theories have an explanation for this dark energy. This is undoubtedly one of the most fundamental questions of today. The origin of the Universe remained outside the premise of science till quite recently. A scientific pursuit of this question during the past 40 years has astonishingly led us into the realm of the unknown!

It is time to conclude this overview. To repeat what I said in the beginning, the main purpose was to give you an overall feeling for the present revolution in astronomy. We shall, in due course, discuss all the topics mentioned above.

But let us now begin. And what better place to begin than the beginning itself. The rest of this volume is devoted to a discussion of the nature of stars, for this is where the subject of astrophysics began.

Chapter 1
What are the Stars?

Historical Introduction

What are the stars?

The splendour of the night sky has fascinated mankind since time immemorial. To the ancient Greeks, the sky was a dark shield pierced with numerous holes through which an outer fire shone. These shining holes were the stars to them. The true nature of the stars remained a mystery for two millennia. Obviously influenced by the ancient Greeks, the great astronomer Johannes Kepler thought that all the stars were roughly at the same distance from Earth and were packed into a thin spherical shell. An important philosophical question at the time of Isaac Newton was the following: '*Is the Sun a star?*' In a characteristic fashion, Newton summarily dismissed this question and asserted that our Sun was indeed a star. And from the pointlike appearance of the stars, he went on to conclude that the stars must be very far away from the Earth, in comparison with the Sun.

But the nature of the stars remained a great mystery. The philosophers took a different point of view. For example, the positivist philosophers, who greatly influenced European thinking in the eighteenth and nineteenth centuries, asserted that *it was in the nature of things that one could never know what the stars are*.

All that changed dramatically with the discovery made by young Fraunhofer in 1817 (see Fig. 1.1). You will recall that Newton had demonstrated that 'white light' contained light of many wavelengths by sending sunlight through a prism and dispersing it into a spectrum of rainbow colours. In 1802, William Wollaston noticed several dark lines in the solar spectrum, and mistakenly concluded that these lines represented boundaries between the colours. By observing the solar spectrum more carefully and with better-quality prisms, Fraunhofer discovered that there were more than 600 dark lines. He went on to determine their 'positions' in the spectrum and thus deduced the wavelengths at which the dark lines appeared. Fraunhofer went on to discover that the spectrum of the light from bright stars also had these dark lines. These have come to be known as the Fraunhofer lines.

G. Srinivasan, *What are the Stars?* Undergraduate Lecture Notes in Physics,
DOI: 10.1007/978-3-642-45302-1_1, © Springer-Verlag Berlin Heidelberg 2014

Fig. 1.1 A German postage stamp to commemorate the two-hundredth birth anniversary of Fraunhofer. The dark lines in the solar spectrum, carefully sketched by Fraunhofer, have been reproduced. Notice that Fraunhofer has singled out some lines and labelled them. This notation continues to be used

Subsequently, Gustav Kirchoff and Robert Bunsen independently demonstrated in laboratory experiments that such dark lines could be produced in the spectrum of light sources by passing the light through transparent substances. A proper and complete explanation of this phenomenon had to wait till the advent of the quantum theory of matter in the early decades of the last century. Nevertheless, great progress was made in the middle of the nineteenth century in understanding these spectral lines. An example of how such a spectrum looks is shown in Fig. 1.2.

The breakthrough came when the great German physicist Gustav Kirchoff formulated his comprehensive Theory of Radiation in 1859. It is customary to state this theory as three laws of radiation, given below:

First law A luminous opaque body emits radiation of all wavelengths, thus producing a continuous spectrum.

Second law A rarefied luminous gas emits radiation whose spectrum consists of a series of bright lines, sometimes superimposed on a faint continuous spectrum.

Third law If white light from a luminous source is passed through a gas, the gas may absorb certain wavelengths from the incident continuous spectrum so that the intensity at those wavelengths will be missing or diminished, thus producing dark lines.

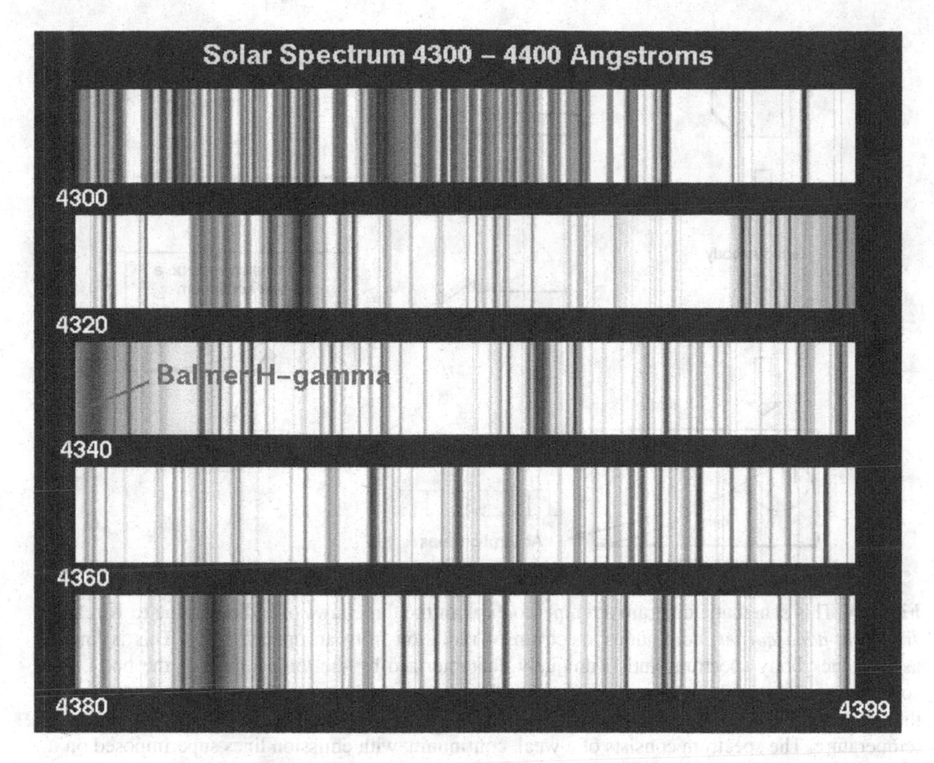

Fig. 1.2 A recent very-high-resolution solar spectrum with many hundreds of absorption lines within a very narrow wavelength interval of just 100 Angstroms

Thus, according to Kirchoff, three kinds of spectra are possible.

1. A continuous spectrum.
2. An emission line spectrum consisting of a series of bright lines.
3. An absorption line spectrum consisting of a series of dark lines.

The profound significance of these laws due to Kirchoff is that each element (or compound) emits or absorbs radiation at specific wavelengths, which are characteristic of that particular element. A schematic is shown in Fig. 1.3.

Consequently, the presence in the spectrum of a particular pattern of bright (or dark) lines characteristic of an element is clear evidence of the presence of that particular element in the source (in the case of bright lines) or between the source and the observer (in the case of dark lines). Figures 1.4 and 1.5 show some spectral lines of hydrogen and iron, respectively. By the end of the nineteenth century, thanks to painstaking laboratory experiments on the elements known at that time, the thousands of Fraunhofer lines in the spectrum of the Sun and the stars were identified with the forty or so elements which had been studied.

A great scientific revolution had started! A question which was considered meaningless, or even frivolous, suddenly acquired a meaning. Contrary to what the

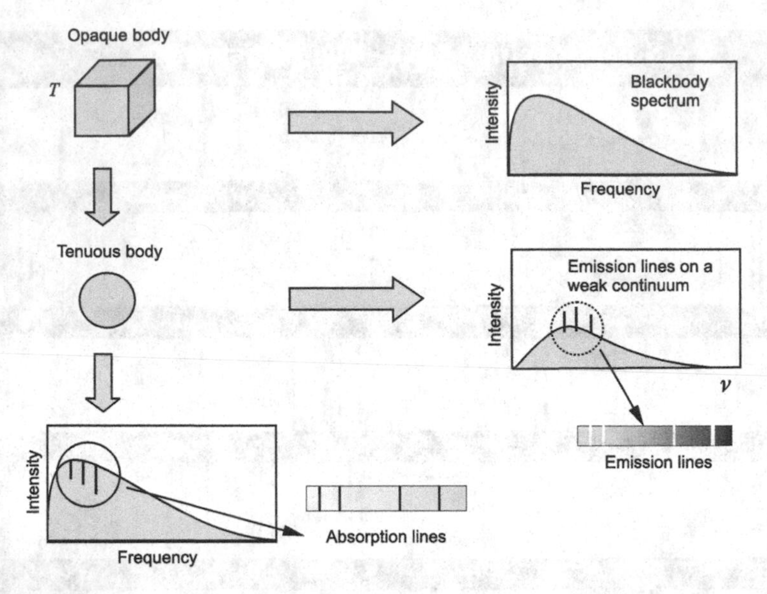

Fig. 1.3 This schematic diagram attempts to explain the three laws of radiation due to Kirchoff. *Top right-hand corner*: continuous spectrum of radiation from an opaque body. This is known as the black body spectrum and is uniquely characterized by the temperature of the body. The spectrum peaks sharply at a frequency which depends upon the temperature of the body. Below that is shown the spectrum of radiation from a transparent body (such as a blob of gas) at a finite temperature. The spectrum consists of a weak continuum, with emission lines superimposed on it. Atoms of each element in the gas produce a unique set of emission lines at specific frequencies that are characteristic of that element. If the radiation from an opaque body passes through a tenuous medium, then dark lines appear in the continuous spectrum. These are caused by atoms in the gas absorbing radiation from the background continuum, partly or wholly, at frequencies at which they are capable of radiating. Thus, the absorption lines caused by atoms of an element will occur at the same frequencies as the emission lines produced by that element. This is shown at the bottom left

Fig. 1.4 Four emission lines from hydrogen in the Balmer Series

Fig. 1.5 Some emission lines from iron

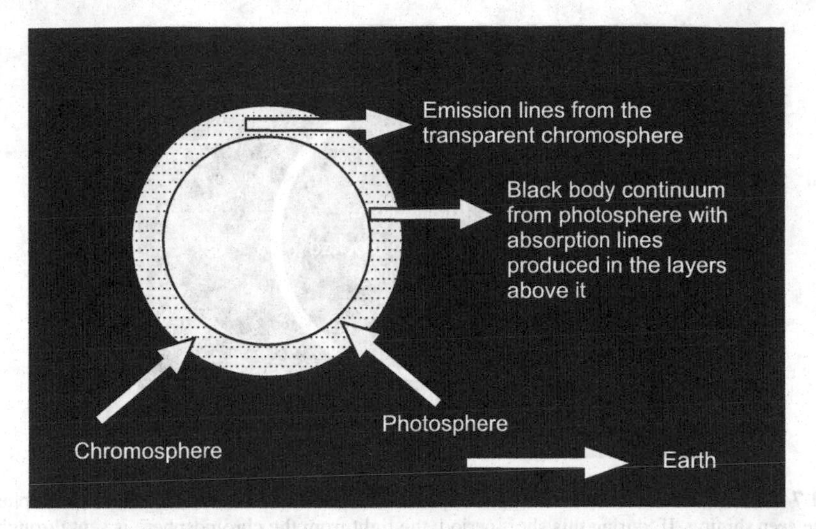

Fig. 1.6 Although the Sun appears like a disc, it does not have a sharp boundary. What appears like a sharp surface is called the *photosphere* (see the text for the meaning of this concept). Above that there is a thin layer which is called the *chromosphere*. This schematic diagram tells what we expect to see when we look at the Sun from the Earth. If we could look at the chromosphere in the direction of the limb, then we expect to see emission lines from the atoms present there; we expect to see this layer in emission because in that direction there is no background radiation. If, on the other hand, we look at the disc of the Sun we are looking at this layer against a strong background radiation from the photosphere. Therefore, we expect to see a strong blackbody continuum from the photosphere, with absorption lines produced by the atoms in the overlying layer. Unfortunately, we cannot see the chromosphere under normal circumstances since the light from it is very faint compared to that from the photosphere—except during a total solar eclipse when the photosphere is blocked by the Moon

philosophers had maintained, we now knew what the stars were. Fraunhofer's discovery and the explanation by Kirchoff had shown us that at least the outer layers of the Sun, and the stars, were gaseous and made of the same elements which are found on the Earth. Moreover, the inner regions of the Sun must be opaque. We know this because the Sun emits a continuous spectrum of radiation (Kirchoff's first law). A deeper understanding of this had to wait until the development of the thermodynamics of radiation, which eventually happened in the final decade of the nineteenth century. We shall return to this a little later.

If the above explanation for the dark lines in the spectrum of the Sun is correct, then there is a clear prediction one could make, which is, if one could detect the radiation emanating solely from the outer layer of the Sun—without any background radiation—then its spectrum should consist of a series of bright lines. Also, that the wavelengths of these bright lines should precisely correspond to those of the dark. Fraunhofer lines (see Fig. 1.6). The question is how to view the outer layers by themselves. When we look at the Sun, we see much of the outer layers in projection in the foreground against the deeper opaque layers. This is the case everywhere in the disc of the Sun, except in the limb of the Sun; there we see only the outer layers.

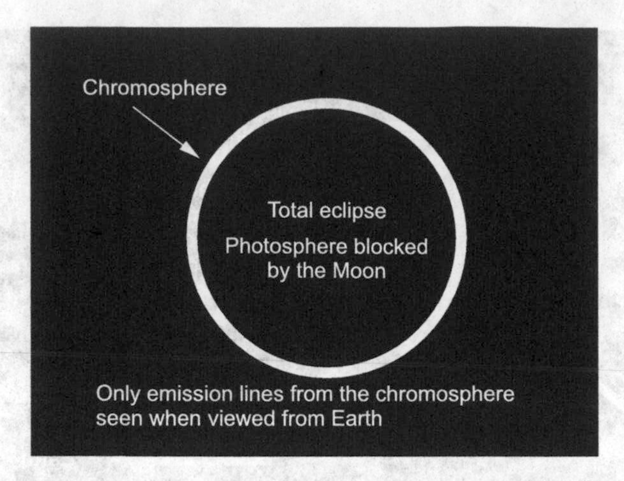

Fig. 1.7 During a total solar eclipse the photosphere is blocked by the Moon, and the chromosphere can be seen shining. If, during this short period, the light from the chromosphere is sent through a spectrograph, then we expect to see bright emission lines. This is exactly what was observed during the eclipse of 18 August, 1868 in Guntur, India. Since this spectacle lasts only for a very short time, such a spectrum is known as the flash spectrum. As explained in the text, it is the analysis of the emission lines in the flash spectrum that led to the discovery of helium

Unfortunately, the disc of the Sun is so bright that the light from it dominates over the light from the limb. Therefore, it is practically difficult to separate the light from the outer layers in the limb of the Sun. So the prediction that the spectrum of light from the limb should consist of emission lines rather than absorption lines, cannot be verified easily. A total solar eclipse, during which the Moon temporarily covers the disc of the Sun, offers a unique opportunity to study the spectrum of the radiation from the luminous, presumably tenuous, atmosphere of the Sun.

Precisely such an observation was made during the total solar eclipse of 18 August 1868 in Guntur in the state of Andhra Pradesh in India. There were several international teams that had set up camp there. The leader of one camp was the well-known French astronomer Pierre Jules Janssen. Janssen and his colleagues found that during the brief moment of totality the spectrum of light from the chromosphere showed a series of discrete emission lines. They termed it the *flash spectrum*, since it lasted only for a few seconds. As the Moon moved across the disc of the Sun, the light from the disc once again dominated over that from the limb; the emission lines disappeared and the Fraunhofer absorption lines were seen once again. *What was remarkable was that the emission lines in the flash spectrum corresponded to the elements whose presence in the outer layers had been inferred earlier in a study of the Fraunhofer absorption lines*. This was a spectacular confirmation of the validity of the laws of radiation enunciated by Kirchoff (see Fig. 1.7).

But there was a surprise! Janssen noticed that the flash spectrum contained a very bright yellow emission line (with a wavelength of 5874.9 Å) which could not be identified with any known element on Earth. This line was so bright that Janssen could see it even after the eclipse was over by carefully placing the slit of his spectrograph at the same location in the limb from where he had seen the yellow line. Two months later, the famous English astronomer Sir Norman Lockyer independently discovered the same bright-yellow emission line. Lockyer was convinced that this line was caused by a new element in the Sun. He was so confident that he named this new element helium (deriving from *helios*, the Greek name for the Sun). It turned out that this element, too, is present on the Earth. In 1895, nearly thirty years after this momentous discovery by Janssen and Lockyer, the famous British chemist Sir William Ramsay identified helium trapped in mineral samples on Earth.

The discovery of helium was very important from another perspective. When Dmitri Mendeleyev prepared the famous *Periodic Table of Elements* in 1869, there was a missing element of atomic number 2. The discovery of helium filled this gap! Astronomy had made yet another very important contribution to physics. Sir William Ramsay went on to discover other rare gases and was awarded the *Nobel Prize for Chemistry, in 1904*. It may interest you to know that in the same year the *Nobel Prize for Physics* was awarded to Lord Rayleigh for his discovery of *argon*, a rare gas, and determining its properties. This may seem quite extraordinary because Lord Rayleigh is generally known for his immensely mathematical and definitive treatment of many complex problems in theoretical physics, but like James Clerk Maxwell, and Enrico Fermi in the twentieth century, Rayleigh was not only a theorist of the highest order but also a great experimentalist!

Let us now return to the implication of all this for understanding the structure of the Sun and the stars. We have argued that the dark lines observed by Fraunhofer in the spectrum of the Sun could be understood if there is a gaseous outer layer. The idea is that the atoms of the elements present in this gaseous layer selectively absorb the continuous spectrum emanating from the interior. According to a fundamental principle of physics, this selective absorption can only happen if the inner region, from which the continuous spectrum originates, *is hotter than* the outer layers, but this is precisely what one would expect on very general grounds. For example, we know that energy is flowing out of the Sun. The laws of thermodynamics tell us that heat flows from a hotter to a cooler body. One may, therefore, expect a gradient of temperature, with the temperature rising as we go into the Sun. There is another observational fact that confirms this. This is known as *limb darkening of the Sun*.

If you look at a good-quality image of the Sun taken with white light, you will notice that the visible disc of the Sun does not appear uniformly bright. The edge of the disc is less intense than the centre of the disc. This is the phenomenon of *limb darkening* (Fig. 1.8).

The Photosphere

Before we try to understand limb darkening of the Sun, there is an embarrassing question we have to address. Planets have sharp boundaries because they are solid. Since the Sun has a gaseous atmosphere, why do we see a sharp boundary? Why does the contour of the Sun appear like a disc? This is not difficult to understand. As we

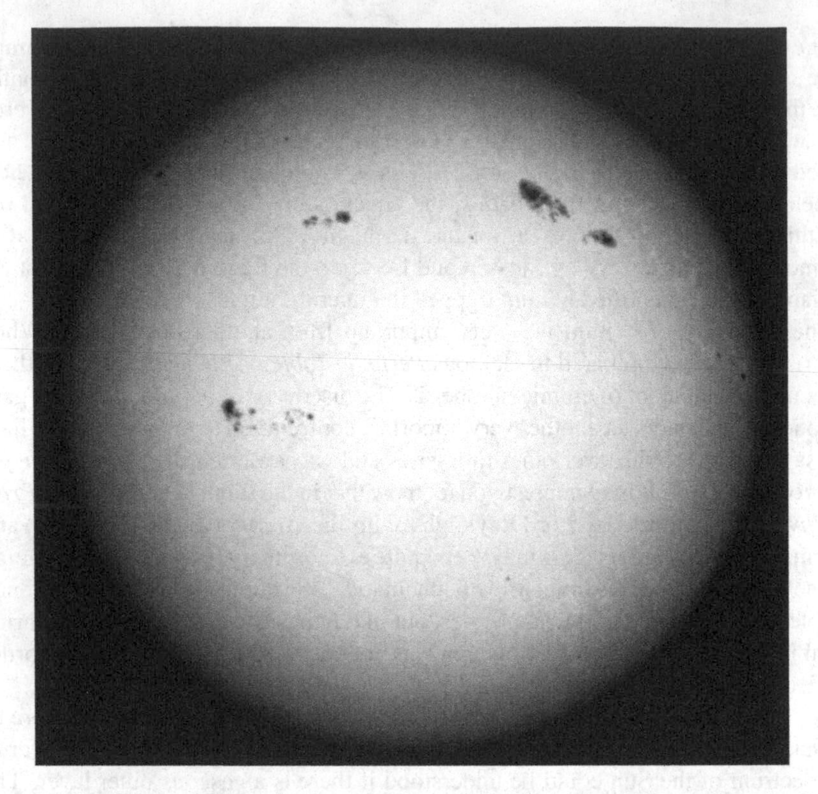

Fig. 1.8 An image of the Sun taken in white light. Notice that the central portion of the disc is brighter than the outer region, known as the limb. This is known as limb darkening. Some prominent sun-spots are also seen. Notice that the Sun-spots occur in clusters

will soon argue, the energy radiated by the Sun originates near its very centre. When it is given out as a byproduct of nuclear reactions, the radiation is mostly in the form of X-rays and gamma rays. On their way out, these *photons* (or quantized bundles of energy) get kicked around quite a bit by electrons and ionized heavy atoms; they are scattered, absorbed and reemitted. Because of this, they cannot stream out to the surface and fly to infinity at the speed of light. Their journey to the surface is a painstaking *random walk* (very much like a drunken man).

It turn out that inside a star like the Sun, the *mean free path* of the photons is only about 0.5 cm (we shall formally define this concept in Chap. 3, *Eddington's Theory of the Stars*, but for now, think of the mean free path as the mean distance travelled by a photon between two successive collisions with matter). With each collision, the direction of the photon is changed. To appreciate how random this walk of the photons is, consider this (see Fig. 1.9). Imagine you embark on a journey to the centre of the Sun. The scenery would be most uninteresting. First of all, you and I would find the interior very dark (!) since there would hardly be any photons whose wavelengths lie in the visible region of the electromagnetic spectrum. As we shall soon argue,

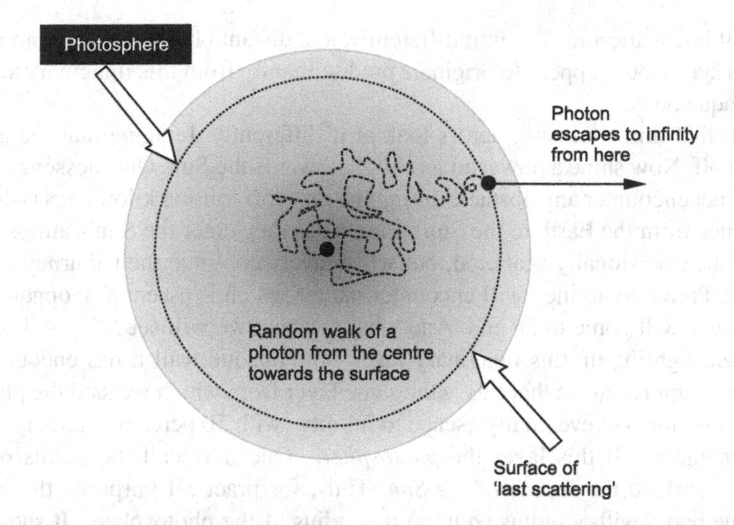

Fig. 1.9 This sketch depicts the laborious journey of a photon from the centre of the Sun, where it is created, to the photosphere. In the absence of any scattering, it would take the photon only about two seconds to reach the surface. Instead, it takes about 30,000 years; for more massive stars, which are bigger, this escape time could be *several million years*! This is because the mean free path of the photons is only about half a centimetre. This means that after travelling a mere 0.5 cm the direction of the photon has changed quite randomly. As the photon diffuses outwards ever so slowly, its energy also has been degraded. This is why the Sun appears brightest at visible wavelengths, even though deep in its interior most of the radiant energy is in the form of X-rays

the average temperature of the interior of the Sun is about ten million degrees. As a consequence, the radiation in the interior mainly consists of X-rays.

You may think that *Superman*—who has X-ray vision—could hope to enjoy the scenery. But even he would be disappointed. He would find that it is like walking in a very thick fog; he would not be able to see even the tip of his nose (photons scattered off the tip of his nose are unlikely to reach his eye since the mean free path is about 0.5 cm)! During this random walk, the energy of the photons also gradually gets degraded until by the time the photon is ready to escape to infinity, the majority of photons will have energy corresponding to the visible wavelengths (*this is why the Sun, as seen from the outside, is brightest in visible wavelengths*). By this time, more than *thirty thousand years* would have elapsed since the photons began their journey from the centre of the Sun. In the case of stars more massive than the Sun (which would also be larger in size) this escape time could be *several million years*!

As the photons reach the outermost region, the ambient density would have decreased, and the matter would be very tenuous. Consequently the mean distance travelled by the photons between two successive collisions would increase; this is a direct result of the mean distance between the scattering particles increasing. Eventually, *the photons would reach a layer from which there is approximately 50 percent chance that they will escape to infinity without further scattering*. In technical jargon, one says that *the optical depth above that surface is unity*. One may call this layer the surface of last scattering. To put it differently, to a distant observer, the photons that reach his eyes would appear to originate predominantly from this imaginary surface of an opaque body.

If you find this confusing, let us look at it differently. Imagine that the Sun is switched off. Now shine a powerful torchlight towards the Sun. Our messenger photons will not encounter any obstacle during the first 150-million kilometres (which is the distance from the Earth to the Sun). Finally, as they enter the Sun's atmosphere, they will be occasionally scattered, but will bravely continue their journey toward its centre. Pretty soon, they will encounter matter which is essentially opaque, and their journey will come to an end. And what is it that we will see? We will see the torch beam lighting up this imaginary layer—the opaque wall it has encountered. The point to appreciate is that this is the same layer from which we said the photons from the interior will eventually escape to infinity (with 50 percent chance).

Astronomers call this layer the *photosphere*. One may call the radius of this imaginary surface the *radius of the Sun*. Thus, for practical purposes the Sun is an opaque body, with a radius equal to the radius of the photosphere. It should be remembered that in reality even the photosphere is not a sharp boundary (unlike the surface of the Earth). It is a layer from which the photons have a 50 percent chance of escaping to infinity. Clearly, the layer from which the photons have a 50 percent chance of escaping depends also on the direction of the photon with respect to the local normal (or vertical). The important thing is that the thickness of this layer—which we call the photosphere—is very small compared to the size of the Sun. It is also important to remember that the *radius* of the Sun (or more correctly, the radius of the photosphere) depends on the wavelength at which you are observing the Sun. Shorter wavelengths would be able to escape from deeper down, while longer wavelengths would have to wait till they diffuse out a little further before they can escape. This is because what matters is a comparison between the wavelength of the radiation and the interparticle distance. For example, if you look at the image of the Sun produced by a radio telescope operating at a wavelength of, say, ten metres, you will find that the Sun is roughly twice as big as it appears in the visible!

Let us now return to the phenomenon of *limb darkening*. Figure 1.10 gives a schematic illustration of this. Remember what I said before: the photons that reach us interacted last with matter at a depth such that the photons have approximately 50 percent chance of escaping without further scattering. The figure shows two lines of sight: one looking towards the centre of the disc and the other towards the limb. Because the line of sight from the limb is at *an angle to the vertical*, it passes through more solar material.

Therefore, for the photons emerging from the limb *at an angle to the local vertical*, the *surface of last scattering* is higher in the atmosphere than for the photons in the line of sight from the centre of the disc. *The phenomenon of limb darkening can now be understood if the higher layers are cooler than the lower layers.* This follows from one of the fundamental laws discovered by Kirchoff (see Fig. 1.10). According to this law, the intensity of radiation from an opaque body is described by a universal function of the temperature of the body: *a hotter body emits more radiation at every*

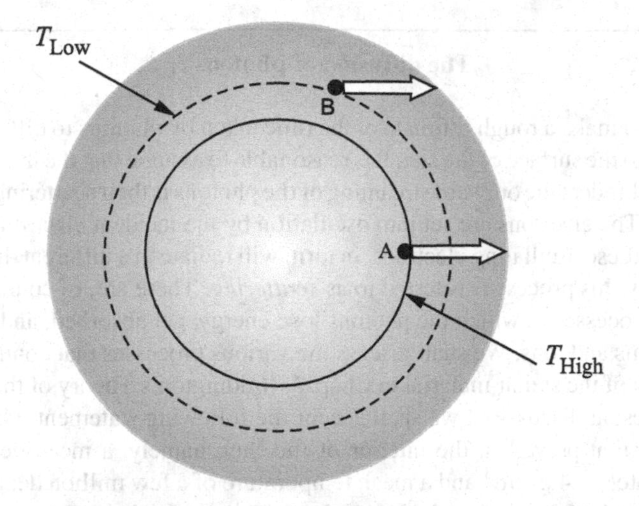

Fig. 1.10 The limb darkening of the Sun tells us immediately that there is a temperature gradient, with the temperature increasing as we go deeper into the Sun. The radiation we receive from any part of the Sun essentially comes from a layer from which there is roughly 50 percent chance for the photon to escape to infinity; we called this layer the photosphere. The radius of this *photosphere* is determined by how much matter lies above it; the effective thickness of the column of matter above it should be low enough so that there is a 50 percent chance for the photon to escape in a *particular direction*. Let A be the point from which the radiation from the centre of the disc reaches us. It is clear from the geometry that the radiation from the limb will only be able to escape from a point B at a larger radius; since there is more matter in the oblique direction, the photons will be able to escape only from further out. The darkened limb tells us that the intensity of radiation from B is less than that from A. This, in turn, tells us that the larger photosphere is at a lower temperature than the smaller photosphere; this follows from Kirchoff's law. Thus, limb darkening is a clear signature of the temperature gradient

wavelength. So the simple fact that the limb of the Sun is darker immediately tells us that the temperature increases as we go into the Sun.

The Interior of the Sun

So far we have been discussing the atmosphere and the outer layers of the Sun. Let us not linger over this any longer. This is not because this region is devoid of interesting features. On the contrary, many exciting phenomena occur here, and the Sun is close enough that we can study these in detail. Curiously, some of these phenomena tell us in detail about what is happening deep within the interior! We shall return to these recent developments in Chap. 6, *Sounds of the Sun*. But for now, our interest is in the interior. Let us plunge into it.

The diffusion of photons

Let us try to make a rough estimate of the time taken by photons to diffuse from the centre to the surface of the star. It is reasonable to assume that the main mechanism that hinders the outward streaming of the photons is their scattering by free electrons. The electrons are set into oscillation by the incident electromagnetic wave, and these oscillating electrons, in turn, will radiate in a different direction. This is why this process is referred to as *scattering*. There are, of course, other physical processes in which the photons lose energy, get absorbed, and reemitted by atoms and ions. We shall discuss the various processes that contribute to the opacity of the stellar material in Chap. 3, 'Eddington's Theory of the Stars'. For the present discussion, we shall accept the following statement. Under the conditions that prevail in the interior of the Sun, namely, a mean density of approximately $1.4\ \mathrm{g/cm^3}$ and a mean temperature of a few million degrees, the mean free path of the photons is about 0.5 cm. This means that a photon will randomly change its direction after travelling for about 0.5 cm. Given this random walk of the photons, with each step of length being approximately 0.5 cm, we wish to know how long it will take for the photon to reach the surface. Remember that between any two successive collisions the photon will be travelling at the speed of light, denoted by the symbol, c. Therefore, the question boils down to the following: after how many collisions will our photon which originated at the centre travel an effective distance equal to the radius of the star and reach the surface?

This is the famous *random walk* problem. Albert Einstein discussed this problem in his epochmaking paper on Brownian motion published in 1905. As you may know, the same year, he published two more incredible papers that changed the course of physics. One of them was on the *Special Theory of Relativity*, and the other on the Photoelectric Effect in which he introduced the revolutionary idea that energy of light is *quantized*.

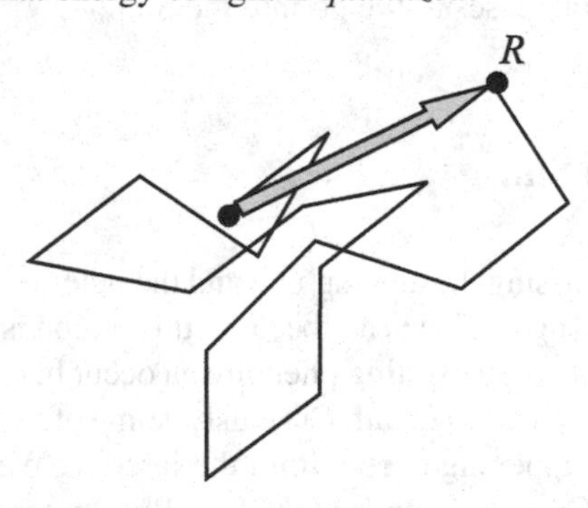

Consider the random walk of a particle in three dimensions. Let the particle commence its walk at the origin and let the step length be denoted by l. The particle changes its direction after every step. Although the particle goes back and forth in a random fashion, it nevertheless slowly wanders further and further from where it started (the famous analogy of a drunken man trying to reach home is often cited). After a certain number of steps, N, it reaches a distance R from the origin. The distance it wanders, R, and the average distance per step, l, are related by the following famous relation:

$$\boxed{R^2 = \frac{1}{3}Nl^2}$$

This relation was first obtained by Einstein. Clearly, the number of steps needed to reach a given distance will depend upon the number of dimensions available for the walk. Think about this.

Let us now return to our photon which is struggling to find its way to the surface of the Sun. The effective distance it has to travel is the radius of the Sun, $R_\odot$. It follows from Einstein's formula that the number of *steps* it has to take is $N = 3R_\odot^2/l^2$, where l is the mean free path of the photons. During these N steps, the actual *linear distance* travelled by the photon is Nl (the number of steps multiplied by the length of the step). Remember that between collisions the photon travels at the speed of light. Therefore, the time taken by the photon to diffuse out to a distance $R_\odot$ is

$$t_{\text{diffusion}} \sim \frac{Nl}{c} \sim \frac{3R_\odot^2}{lc}.$$

We can now substitute the value for the radius of the star and the velocity of light. Recall that the mean free path of the photon in the stellar interior is approximately 0.5 cm. For the Sun, this diffusion time is approximately *30,000 years*. A star Which is ten times as massive as the Sun will be approximately ten times bigger. Therefore the photon diffusion time will be a hundred times larger, approximately, *several million years!* (You may verify these time scales by substituting the given numerical values).

Let us first recall some of the basic facts. The mass of the Sun (in grams) is $M_\odot = 2 \times 10^{33}$ g. Its radius is approximately a million kilometres (6.96×10^{10} cm). Dividing the mass by its volume, we get a mean density of 1.4 g/cm^3. This is just a little more than the density of water. The mean density of some of the stars is much less than this. For example, the mean density of the giant star Capella is nearly the same as the density of air. Is the interior of the Sun solid, liquid or gaseous? The famous eighteenth-century astronomer William Herschel thought that the Sun was a cold solid body, surrounded by glowing clouds of gas (from which the radiation we

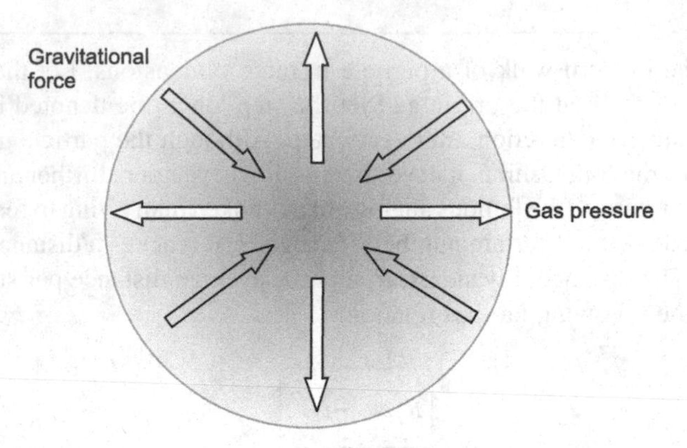

Fig. 1.11 A star is stable because the inward directed force due to self gravity is opposed, and balanced, by the pressure of the gas. The weight of the column of gas above any point in the star must be countered by the pressure of the gas. This condition must be satisfied at every point in the star. Otherwise the star will not be in mechanical or hydrostatic equilibrium. A precise mathematical statement of this condition will be set up in Chap. 3, *Eddington's Theory of the Stars*

receive emanates), but as we shall argue below, because of the enormous pressure and temperature that obtains inside the Sun, the matter in the interior must also be in a gaseous form.

Using standard mathematical methods, it is possible to calculate how fast the pressure increases as we go into the Sun, and how fast the temperature must increase. The two are related; indeed, they must be. If the Sun is a hot blob of gas, then it should expand in all directions and disperse. The Sun is stable because it is held together by its own gravity; every part of the star attracting every other part according to Newton's law. But there is a problem; if gravity was not opposed by some other force, then the star will collapse due to the attractive nature of the gravitational force. What saves it is the pressure exerted by the gas. Imagine yourself at some point inside the star. As may be seen in Fig. 1.11, the weight of all the layers above you will try to push you down. This is opposed by the tendency of the gas below you to expand and push you up. To put it differently, gravity is opposed by the pressure of the gas. In a steady state, these two tendencies must be precisely balanced, and this must be so at every point in the star.

Let us try to understand this differently. In classical physics, a gas exerts pressure because at any finite temperature the constituent particles of the gas are in a perpetual state of motion. Not all of them have the same velocity. While it is not possible to define the velocity of any particular particle, one can make a *statistical statement* about the probability of the number of particles having a velocity in any narrow interval. This probability distribution of velocities was discovered by James Clerk Maxwell, and was one of the great intellectual achievements in physics in the nineteenth century. We shall not digress to discuss this achievement here, but return to it a little later. For the purpose of the present discussion we shall merely note the following. Although the individual particles have different velocities, and

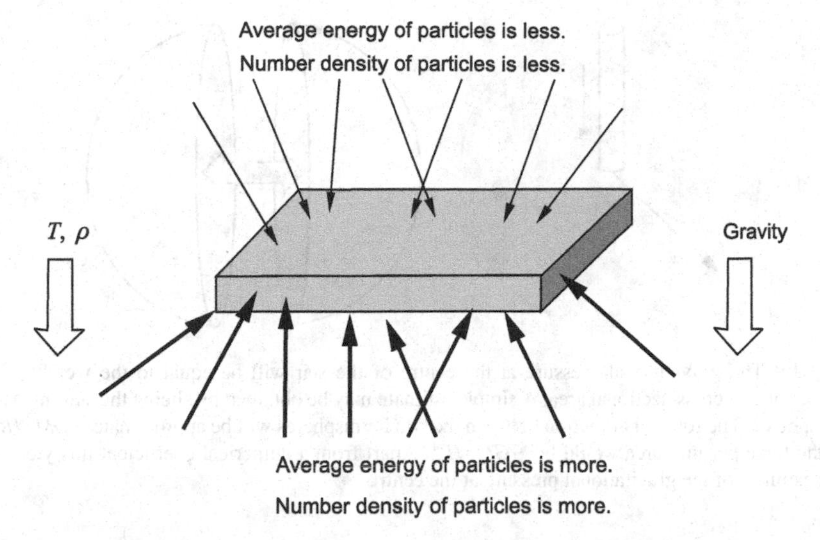
Average energy of particles is less.
Number density of particles is less.
T, ρ
Gravity
Average energy of particles is more.
Number density of particles is more.

Fig. 1.12 Imagine an infinitesimal slab of the stellar material at some radial distance from the centre of the star. Left to itself, this slab will sink due the gravitational attraction of matter which is interior to the slab in question. For it to be stable, it must be supported by a net outward force arising from the collision of particles on the two sides of the slab; each collision results in a transfer of momentum to the slab. Remember that both the temperature and the density increase as one goes deeper into the star. The higher temperature implies that the average energy of the particles colliding on the lower side will be more. The higher density implies that there will be more collisions per unit time on the lower surface. Consequently, the pressure on the lower side will be more than on the upper side. The star will have to adjust its temperature gradient, as well as the density gradient, such that the pressure differential precisely balances the gravitational force (per unit area). This condition has to be satisfied at every point in the star

therefore different energies, it is possible to calculate the *most probable speed* and the *most probable energy* of the particles. For a gas confined in a three-dimensional box, the most probable or *average energy* of the particles is $3/2\, k_B T$, where T is the temperature of the gas and k_B is Boltzman's constant. As you will notice, the *hotter the gas is, greater is the average energy of the particles*. As these particle collide against the walls of the container they will be reflected back, and in the process, they will transfer momentum to the walls. In other words, they will exert a force on the walls. Pressure is just this force *per unit area*. Naturally, this force (or pressure) will depend not only on how hot the gas is, but also on the number of particles per unit volume.

Let us return to our star, and focus attention on a small piece of the stellar material at some distance from the centre of the star. Our concern is how the matter above it is supported against gravity. As shown in Fig. 1.12, in the microscopic picture, the support is provided by incessant collisions by particles underneath. This is the same way a car is supported by its tyres. I leave it to you to understand this delicate balance between the inward directed force of gravity and the pressure exerted by the gas. At any point in the star, if we could estimate the weight per unit area of the material above

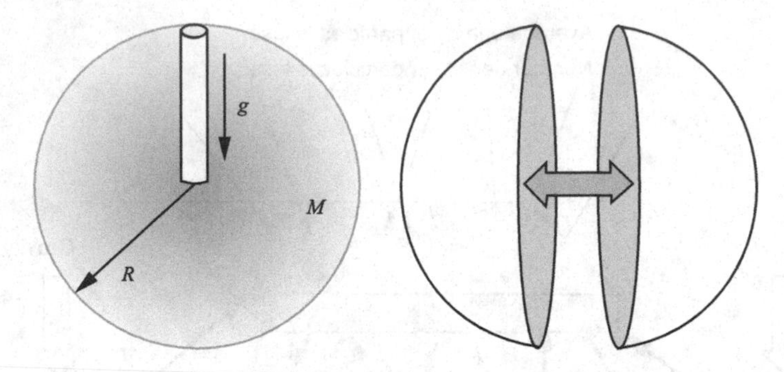

Fig. 1.13 The gravitational pressure at the centre of the star will be equal to the weight of a cylinder of unit cross-sectional area. A simple estimate may be obtained by slicing the star into two hemispheres. The force of attraction between the two hemispheres will be approximately GM^2/R^2, and the force per unit area would be $\sim GM^2/R^4$. Apart from a numerical coefficient this yields a good estimate of the gravitational pressure at the centre

it (the gravitational pressure, if you like) then we would know the pressure the gas must exert to support this weight, but this will not rightaway tell us the temperature at that point. This is because the pressure exerted by a gas depends on both the temperature and the density. This combination cannot be arbitrarily assigned, but has to be solved for, in a self-consistent manner.

We shall set up the necessary equations presently, but let us first make a crude estimate of the gravitational pressure near the centre of the Sun. An accurate calculation is laborious, but one can make a simple-minded estimate. We will proceed along lines similar to how one will estimate the Earth's atmospheric pressure at sea level. The pressure of the atmosphere is the weight of the air mass per unit area above the sea level. In a similar way, the gravitational pressure (or the gravitational force per unit area) at the centre of the Sun is the weight per unit area of the stellar material on top of it.

$$P_c = \mu g,$$

where μ is the mass per unit area, and g is the average acceleration due to gravity felt by the column of material. The mass per unit area must be of the order of $M_{\odot}/R_{\odot}^2$ (Imagine slicing a sphere into two halves, as shown in Fig. 1.13. Place one of the hemispheres on the table. The mass per unit area at base of the hemisphere will, of course, depend on the location of the unit area at the base of the hemisphere, but it has to be of the order of M/R^2. There will be numerical coefficient multiplying this, which can be determined if one is willing to do an honest calculation using calculus!). In a similar way, the average acceleration due to gravity must be some fraction of the acceleration at the surface, viz., GM/R^2. Therefore, the gravitational pressure at the centre must be roughly

$$P_c \sim G\frac{M_\odot^2}{R_\odot^4}, \qquad (1.1)$$
$$\propto M^2,$$
$$\propto \frac{1}{R^4}.$$

If you are not happy with this estimate, try thinking of it this way. Again, imagine slicing a sphere into two hemispheres and separate them slightly. Apart from a numerical coefficient, the force of attraction between the two hemispheres must be of the order GM^2/R^2. Therefore, the force per unit area must be approximately $\sim GM^2/R^4$. A more proper calculation gives a numerical coefficient of about 20, but we have not done too badly, given how simple minded our estimate was! What is of interest is the numerical value of this pressure. Substituting for the mass and radius of the Sun we obtain a value approximately equal to 2×10^{17} g cm^{-1}s^{-2}. The pressure at the centre of the Sun is truly enormous; it is nearly a million million times the atmospheric pressure at sea level on Earth!

What can we say about the central temperature? You will recall from our earlier discussion that gas pressure at the centre must equal the gravitational pressure. To estimate the central temperature we must know the central density. A detailed mathematical calculation gives a central density which is approximately 150 g cm^{-3} (about 110 times the average density of 1.4 g cm^{-3}). This yields a central temperature of *15 million degrees kelvin*!

It is perhaps more useful to have a feel for the average temperature of the Sun. This can be estimated in a much more elegant and satisfactory manner. To do this, we shall use a very powerful theorem known as the *Virial Theorem*.

The Virial Theorem

This theorem is of very general validity, and is applicable as long as the system under consideration is statistically stable. This applies, for example, to a planet orbiting the Sun. The theorem states that in the steady state *the total energy* of the system is equal to *one-half the potential energy*. For a system like the Earth orbiting the Sun, the total energy is the sum of the kinetic energy of the Earth (due to its motion in the orbit) and the gravitational potential energy of the Earth (due to it being attracted by the Sun). The theorem states that

$$\frac{1}{2}mv^2 + \left(-\frac{GM_\odot m}{r}\right) = \frac{1}{2}\left(-\frac{GM_\odot m}{r}\right). \qquad (1.2)$$

The first term on the left-hand side is the kinetic energy of the Earth as it revolves around the Sun. The second term is the gravitational potential energy of the Earth at a distance from the centre of the Sun. According to the virial theorem, the sum of the kinetic energy and the potential energy of the Earth must be equal to *one-half* the potential energy. This relation can be proved rather simply. You will remember that for the Earth to be in a stable orbit around the Sun, the centrifugal force acting on it must be balanced by the gravitational force.

$$\frac{mv^2}{r} = \frac{GM_{\odot}m}{r^2}.$$

Using this relation, Eq. (1.2) is easily verified.

This powerful virial theorem can be invoked to estimate the average temperature of the Sun. In this case, the total energy is the sum of the stored thermal energy in the Sun and the *gravitational potential energy* of the Sun due to self-attraction. According to the virial theorem,

$$\text{Thermal energy} + \text{Grav. potential energy} = \frac{1}{2}\,\text{Grav. potential energy}.$$

Therefore,

$$\boxed{\text{Thermal energy} = -\frac{1}{2}\,\text{Grav. potential energy}.}$$

(Notice the minus sign on the right-hand side. Since the gravitational force is attractive, gravitational potential energy will be negative. So the minus sign is needed to make the right-hand side positive). The gravitational potential energy of a sphere of mass M and radius R is $\sim -GM^2/R$ (if you know some calculus, try deriving this). The thermal energy of the Sun is just the sum of the kinetic energy of the constituent particles. Let T be the average temperature of the Sun. We remarked above that the *average energy of the particles* is $3/2\ k_BT$. If N is the total number of independent particles then the total thermal energy is $3/2\ Nk_BT$. Thus, according to the virial theorem,

$$\frac{3}{2}Nk_BT = \frac{1}{2}G\frac{M^2}{R}. \tag{1.3}$$

Since we know the mass of the Sun, we can estimate the number of particles by assuming certain chemical composition. The above equation can then be solved for the average temperature. This yields a value of *10 million degrees kelvin*. (Take a few minutes to verify this. Since you know the mass of the Sun, you can estimate the number of atoms in the Sun. Assume for simplicity that the Sun is made solely of hydrogen.) I hope you are astonished at the power of the virial theorem, which enabled us to make this estimate. Sitting here on Earth, we can say with considerable confidence that the average temperature of the Sun must be ten million degrees! We only needed to know the mass and radius of the Sun. But you may be puzzled by this result. When we look at the Sun, it appears as an opaque body whose *surface temperature* is about 5,800 kelvin. Is this consistent with an average temperature of 10 million kelvin?

There is no problem. What do we mean by saying that the surface temperature of the sun is 5,800 K? What we mean is that the radiation leaving this surface has a temperature of 5,800 K. If we plot the spectrum of this radiation (intensity versus wavelength) it is, to a good approximation, the spectrum of radiation from a black body whose temperature is 5,800 K. Let us recall what Kirchoff taught us.

- In an opaque body, matter and radiation come to true thermodynamic equilibrium.
- The radiation from the surface of the opaque body has lost all memory of the quantum mechanical processes that produced the radiation in the interior, and its original characteristics, such as frequency, direction, polarization, etc.
- The spectrum of this radiation is characterized uniquely by a *universal function* of temperature, and does not depend upon any of the other properties of the body.
- Because matter and radiation were in true thermodynamic equilibrium in the interior, the temperature characterizing the spectrum of the radiation is the same as the *temperature of the matter with which it last interacted.*

So, what we infer from direct observation of the Sun is that the temperature of the *photosphere* is approximately 5,800 K. But this is not in conflict with our estimate for the temperature of the interior. Detailed numerical calculations show that the average temperature of the interior is, indeed, nearly 10 million kelvin. The temperature drops rather rapidly as one approaches the photosphere, to approximately 6,000 kelvin.

Let us now summarize.

Our discussion so far has led us to the following conclusions: The Sun is a gaseous body, held together by its own gravity. The inward-directed gravitation force is balanced by the pressure of the gas. The fact that the Sun is stable tells us that a star settles down at that radius where the temperature of the gas is adequate for gas pressure to support the star against gravitational collapse. The central temperature of the Sun is approximately 15 million kelvin, and the average temperature is approximately 10 million kelvin.

Chapter 2
Stars as Globes of Gas

A Theory of the Stars

J. Homer Lane was the first person to investigate the details of the temperature distribution within a star. In 1870, he published a seminal paper in the *American Journal of Science and Arts*, entitled, 'On the theoretical temperature of the Sun, under the hypothesis of a gaseous mass maintaining its volume by its internal heat, and depending on the laws of gases as known to terrestrial experiment'. Put simply, in this work Lane assumed that stellar matter behaved as an *ideal gas* and obeyed Boyle's law, as terrestrial gases do. It is a different matter altogether whether this is a reasonable assumption or not, and we shall have occasion to return to this in a later chapter. But at the time, this was a significant piece of work, and this pioneering paper was followed by investigations by A. Ritter, in Germany, Lord Kelvin, in Britain, and others. The culmination of this line of investigation was the publication of the monumental book, *Gaskugeln*, by R. Emden, in 1907.

Hydrostatic Equilibrium

Before understanding Lane's results, let us set up the equation for the mechanical stability of the star. Consider an imaginary concentric spherical surface of radius inside the star as shown in Fig. 2.1. Let us place on this surface a small cylinder, whose axis points along the outward radius at that point. The cross-section of this cylinder is of unit area of the base and length dr, and it contains stellar material. The density of this stellar material is $\rho(r)$, the value that obtains at that radius. The gravitational force on that cylinder would be due to the mass interior to the imaginary surface. Let us call this mass, $M(r)$.

As the area of cross-section of the cylinder is unity, and its length is dr, the mass of the infinitesimal cylinder is given by $\rho(r)dr$. The force of attraction between $M(r)$ and $\rho(r)dr$ is

$$\frac{GM(r)\rho(r)dr}{r^2}. \tag{2.1}$$

G. Srinivasan, *What are the Stars?* Undergraduate Lecture Notes in Physics,
DOI: 10.1007/978-3-642-45302-1_2, © Springer-Verlag Berlin Heidelberg 2014

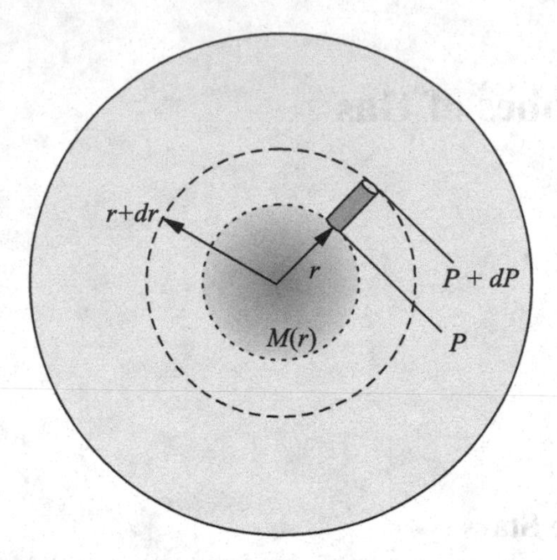

Fig. 2.1 Consider an infinitesimal cylinder at a distance r from the centre of unit cross-sectional area and of height dr. The gravitational force acting on it will arise from the mass $M(r)$ of material interior to the spherical shell on which it lies. This has to be balanced by the difference in pressure dP, which represents a force $-dP$ in the direction of increasing r (pointing outward from the centre). This is the condition for hydrostatic equilibrium of the *star, and must be satisfied at every point in the star*

As you know, in Newton's law the contribution to the force from the mass *exterior* to the surface cancels out. This is illuminating and quite simple to prove. If you know some calculus, I urge you try to prove this. The gravitational force on this infinitesimal cylinder has to be balanced by the pressure differential on it. This is just the *difference* in pressure at the two surfaces of the cylinder at a distance from the centre equal to the radius r and $r + dr$, respectively. Let us denote this by dP. This difference in pressure dP represents the force, $-dP$, acting on the cylinder in the direction of increasing r. Thus the equation for the equilibrium of the unit cylinder is

$$dP = -\frac{GM(r)\rho(r)dr}{r^2}$$

One can rearrange this as

$$\boxed{\frac{dP}{dr} = -\frac{GM(r)\rho(r)}{r^2}}. \tag{2.2}$$

The above equation is known as the *Equation of Hydrostatic Equilibrium*. For a star to be mechanically stable, this equation has to be satisfied at *every point in the star*. Otherwise, as a distinguished astronomer said, 'the punishment would be swift'. Any violation of this condition of *hydrostatic equilibrium* would result in motion within the star. For example, the material within our sample unit cylinder would either sink, or float up due to buoyancy.

Now, to go back to Lane, the pressure of the gas is to be calculated according to Boyle's law, namely,

$$\boxed{P = nk_B T.} \tag{2.3}$$

Here n is the number density of particles (number of constituent particles per unit volume).

The ultimate objective of any theory of the stars is to derive the radial profile of the density and temperature. This is not as simple as it sounds. While the right-hand side of Eq. (2.2) involves only the density, the left-hand side involves both the density and temperature; this is because gas pressure is determined by density, as well as temperature (see Eq. 2.3). Therefore, to derive how the stellar density varies as a function of r one has to know how the temperature varies with r. It should be clear from Eq. (2.3) that dP/dr involves both $d\rho/dr$ and dT/dr. As we shall see in the next chapter, one had to wait till the 1920s to be able to derive the radial profile of the density and temperature.

Why Does the Sun Shine?

Now that one has a mathematical framework to understand why the Sun is stable, the next question to answer is, 'what is the source of energy that makes the Sun shine?' Lane did not attempt to address this question. But his theory did make the following curious assertion. As the star radiates energy, its internal temperature must decrease (since the internal energy is decreasing). This will disturb the delicate balance between the gravitational force and the pressure force (According to thermodynamics, *the pressure of a gas is just two-thirds of its energy density*). Consequently, gravity will gain an upper hand. The star will therefore have no option but to contract. But this works to compress the gas even more, and the gas will get hotter and hotter. *So we have the paradox that as the star radiates energy, it will get hotter!* This is surely a violation of the laws of thermodynamics. As you know, any body that is capable of coming to thermal equilibrium with its surroundings must get *cooler* as it radiates heat to the surrounding. But the above mentioned paradoxical behaviour is what Lane's model predicts, and there is no escaping from this as long as the gas behaves as a perfect gas, and if the energy radiated is coming at the expense of the stored thermal energy. To put it differently, Lane's model star has *negative specific heat*. This discussion naturally leads us to ask the following questions.

The Pressure of an Ideal Gas

Consider an enclosure of volume V containing an ideal gas at a temperature T. Let N be the number of particles in the box. We want to calculate the pressure of the gas. First, what does one mean by an *ideal gas*? What Maxwell taught us is that the particles are in random motion, so that they collide with one another frequently.

Apart from these collisions, the particles can also *interact* with one another through interatomic forces; they may attract or repel one another according to some law. Thus the particles not only have kinetic energy but also potential energy. If the potential energy is negligible compared with the *kinetic energy*, one says that a gas is *ideal*.

$$\text{Kinetic energy} \gg \text{potential energy}$$

A normal gas, such as air, is very nearly ideal because the mean distance between the particles is large compared to the size of the molecules, and this condition is easily satisfied.

According to thermodynamics, the pressure of a gas is equal to two-thirds its energy density (i.e., its internal energy U per unit volume). We have already remarked that the average energy of the particles is $\frac{3}{2}k_BT$. Therefore, the total energy of the gas is $N \times \frac{3}{2}k_BT$. Thus the pressure of the gas is

$$P = \frac{2}{3}\frac{U}{V} = \frac{Nk_BT}{V} = nk_BT, \tag{2.4}$$

where n is the number density of particles. This is *Boyle's law.*

Source of Energy

Why does the Sun shine? And what keeps it shining?

One of the first persons to ponder about this was the English astronomer John Herschel (1792–1871). He was the son of William Herschel, whose name we have already encountered. In 1837, John Herschel exposed a bowl of water to sunshine for a fixed period of time and measured the rise in temperature. This enabled him to estimate the amount of energy radiated by the Sun per unit time. And it was staggering! At first he entertained the idea that the source of energy was some sort of combustion in which chemical energy is converted to heat. But he soon abandoned this idea when he realized that the Sun would not be able to sustain its energy generation for a long time by consuming chemical fuels. He went on to a guess '*If a conjecture might be hazarded, we should look rather to the known possibility of an indefinite generation of heat by friction, or to its excitement by the electric discharge, than to any actual combustion of ponderable fuel, whether solid or gaseous, for the origin of solar radiation*'.

A few years later, in 1846, Julius Mayer (1814–1878), a German physicist, proposed that a continuous bombardment of the Sun by *meteorites* leftover from the time of formation of the solar system, could heat up the Sun. There were grave difficulties with this idea, although initially this idea met with the approval of Lord Kelvin, the high priest of physics at the time.

An alternative explanation was advanced by a Scottish engineer John Waterston. He proposed that gravitational contraction of the Sun at the rate of a hundred metres a year would provide an adequate supply of heat. This idea was picked up by the great German physicist Hermann von Helmholtz since this seemed natural to him. Prior to this, the German philosopher Immanuel Kant had proposed that the solar system formed due to the contraction of a giant cloud of gas (It is incredible that this is, in fact, the modern scenario for the formation of stars and their planets!). Helmholtz felt that this contraction must be continuing still. Lord Kelvin also became convinced of this and abandoned the meteorites hypothesis.

Let us make sure that we are clear about this idea. When a star contracts, matter moves towards the centre of the star; the *difference* in the gravitational potential energy between the old configuration and the new configuration is converted into heat. But there is curious twist to this. Remember what we said earlier. As long as the gas behaves as a *perfect gas* the star must get hotter as it radiates and contracts. So the heat generated as the star contracts must be sufficient not only to replace the heat lost as radiation but also to heat the star to a higher temperature. This is essential, for otherwise the star has no option but to collapse.

Helmholtz and Kelvin estimated that the Sun had been shining for about twenty million years, and will continue to shine for another twenty million years or so. Let us see how one may estimate this timescale. Recall our discussion of the *Virial Theorem* in the previous chapter. According to this theorem, when a star contracts, *only one-half of the gravitational potential energy released is available for radiation*. The other half is stored as thermal energy. The gravitational potential energy of the Sun is $\sim -2GM_\odot^2/R_\odot$. If we divide one-half of this by the rate at which the Sun has been radiating, then we can get an estimate for how long the Sun has been shining.

$$t \sim -\frac{1}{2}\frac{\text{gravitational potential energy}}{\text{luminosity}} \sim \frac{GM_\odot^2/R_\odot^2}{L_\odot} \tag{2.5}$$

This is just like estimating how long the money in your bank account will last; you have to divide the balance in your account by the rate at which you are spending the money. The present rate at which the Sun is losing energy is the *luminosity* of the Sun [$L_\odot = 4 \times 10^{33}$ erg s^{-1}]. Inserting the values for the mass and radius of the Sun, we come to the conclusion that if the Sun had been radiating at the present luminosity, then it could have done so only for about 20 million years [Convince yourself of this by substituting the values]. This seemed a comfortably long time for Lord Kelvin. Even though the geologists were convinced (even at that time) that the Earth was older than 20 million years, Lord Kelvin was not bothered. He used his status to tell the geologists to confine themselves to this timescale!

But he should have been bothered. Most of the stars we see in the sky with the naked eye are far more luminous than the Sun; they radiate a hundred or a thousand times more than the Sun. If he had used the same argument, he would have come to the conclusion that these luminous stars were born only 100,000 years ago. He would then have had to wonder whether, 'the antiquity of man is greater than that of the stars shining', as Eddington put it.

The discovery of radioactivity was the last nail in the coffin for the contraction hypothesis. Using modern techniques, geologists were able to determine the age of the older rocks, and this turned out to be *more than a billion years*. This is how they did it. Uranium is seen to disintegrate into lead and helium at a known rate. Chemically, uranium and lead are very dissimilar; therefore they are unlikely to be deposited together. So if you find both uranium and lead in rocks, you can assume that the lead was formed due to the radioactive disintegration of uranium. Since one knows the rate at which this disintegration takes place, from the ratio of lead-to-uranium one can estimate how long ago the rock formed. Now, if the earth itself is several billion years old, the Sun must be even older than this. So we are back to square one as far as the source of energy in the Sun and the stars. But one can say this. If external source of energy (such as meteorite), as well as gravitational contraction, are ruled out then the *star must contain some hidden source of energy which enables it to shine for billions of years*.

Sir Arthur S. Eddington provided the breakthrough. Addressing the British Association in Cardiff on 24 August 1920, Eddington argued that only *subatomic energy* is available in unlimited quantity. To quote Subrahmanyan Chandrasekhar, 'This address contains some of the most prescient statements in all of astronomical literature'. Eddington's remarks are so bold and incredibly brilliant that I shall quote in full the relevant part of his address.

> Only the inertia of tradition keeps the contraction hypothesis alive—or rather, not alive, but an unburied corpse. But if we decide to inter the corpse, let us frankly recognize the position in which we are left. A star is drawing on a vast reservoir of energy by means unknown to us. This reservoir can scarcely be other than the subatomic energy which, it is known, exists abundantly in all matter; we sometime dream that man will one day learn how to release it and use it for his service. The store is well-nigh inexhaustible, if only it could be tapped. There is sufficient in the Sun to maintain its output of heat for 15 billion years....
>
> Aston has further shown conclusively that the mass of the helium atom is even less than the sum of the masses of the four hydrogen atoms which enter into it—and in this, at any rate, the chemists agree with him. There is a loss of mass in the synthesis amounting to 1 part in 120, the atomic weight of hydrogen being 1.008 and that of helium just 4. I will not dwell on his beautiful proof of this, as you will no doubt be able to hear it from himself. Now mass cannot be annihilated, and the deficit can only represent the mass of the electrical energy set free in the transmutation. We can therefore at once calculate the quantity of energy liberated when helium is made out of hydrogen. If 5 per cent of a star's mass consists initially of hydrogen atoms, which are gradually being combined to form more complex elements, the total heat

> liberated will more than suffice for our demands, and we need look no further for the source of star's energy.
>
> If, indeed, the subatomic energy in the stars is being freely used to maintain their great furnaces, it seems to bring a little nearer to fulfilment our dream of controlling this latent power for the well being of the human race–or for its suicide.
>
> Reproduced from *Observatory,* **43**, 353–5, 1920.

To appreciate how extraordinarily prescient these predictions were we have to transform ourselves to the year 1920. At the time, our knowledge of the atom and its nucleus was very rudimentary. Electrons and protons were the only known elementary particles at that time. Ernest Rutherford had demonstrated that radioactive substances emitted three types of radiation, which he called *alpha, beta* and *gamma radiation*. Beta radiation turned out to be streams of electrons, while *gamma radiation* showed all the characteristics of electromagnetic radiation. By a series of careful experiments, Rutherford was able to demonstrate that the alpha radiation was nothing but the ions of helium atom; they were positively charged with a charge equal to twice the charge of the electron. After this came the brilliant experiments of Rutherford and Soddy. Soon they could draw definite conclusions regarding the atomic mass, or atomic weight, of the alpha particles. The helium atom or alpha particle had four units of atomic weight.

Rutherford then embarked on the now famous *scattering experiments*, in which he bombarded matter with alpha particles. Based on the results of these experiments, Rutherford was able to give a general picture of an atom–a *nuclear atom*. The atom consisted of a minute nucleus, containing practically all the mass of the atom, and a positive charge equal to the atomic number times the magnitude of the electronic charge. For electrical neutrality, the nucleus would have to be surrounded by a number of electrons equal to the atomic number. This deduction of the nuclear nature of the atom was one of the most important discoveries in the entire history of physics. Barely two years later, in 1913, Niels Bohr published his *theory of the hydrogen atom.*

By the year 1920, Rutherford and his brilliant students at the Cavendish Laboratory in Cambridge had succeeded in the artificial disintegration of the atoms. Around the same time, one of his students, Aston, invented the *mass spectrograph*. Using this Aston was able to measure the masses of the atoms. If we take the mass of the hydrogen atom (1.008 units) to be the unit with which to measure the masses of other nuclei, the mass of the helium atom would be exactly 4, and that of oxygen, 16.

This suggested to Eddington that the helium atom must have *formed* by the combination of four hydrogen atoms. But how could this happen? Although the mass of the helium atom made this hypothesis plausible, the question remained, 'is it possible'? How does one account for the fact that the atomic number (or charge) of the helium nucleus is only two? No one knew the answers to these questions, not even Eddington. He simply packed four protons and two electrons into the helium nucleus; this would fix the problem with the charge! Despite all this confusion, he attached supreme significance to Aston's discovery that *the mass of the helium nucleus is less than four times the mass of the hydrogen nucleus*. The energy equivalent of this deficit mass, which we now call the binding energy, is presumably radiated away. According to Eddington, *this is the source of the energy that makes the stars shine.*

Neither the physicists nor the chemists, leave alone the astronomers, thought much of all this. To them, Eddington replied as follows.

> To my mind the **existence** of helium is the best evidence we could desire of the possibility of the **formation** of helium. The four protons and two electrons constituting its nucleus must have been assembled at some time and place; and why not in the stars? When they were assembled the surplus energy must have been released, providing a prolific supply of heat. Prima facie this suggests the interior of a star as a likely locality, since undoubtedly a prolific source of heat is there in operation. I am aware that many critics consider the conditions in the stars not sufficiently extreme to bring about the transmutation—the stars are not hot enough. The critics lay themselves open to an obvious retort; we tell them to go and find a **hotter place**.

Well, once upon a time there was a hotter place! That was the very-early universe. But that is a different story, and we shall discus that much later in our series of monographs. As to how the story of the energy production in the stars unfolded, we shall return to this in Chap. 5. In the meantime, let us go back to Lane's idea and continue with our discussion of the equilibrium and stability of the stars.

We have already discussed the condition for *hydrostatic or mechanical equilibrium:*

$$\frac{dP}{dr} = -\frac{GM(r)\rho}{r^2}. \tag{2.6}$$

To proceed further, we have to supplement this with an equation describing the *thermal equilibrium* of the star. Thermal equilibrium requires that the temperature distribution is capable of maintaining itself automatically even though there is a continuous transfer of heat from one part of the star to another. How does energy in the form of radiant heat flow from the interior to the surface from where it escapes into space? In principle, there are three modes of transfer of heat, viz. *conduction, convection,* and *radiation.* Lane considered convective transport in which heat is transported from one region to another by actual movement of material. This is what happens in our atmosphere. If convective transport is operative, then the star cannot possibly be in mechanical equilibrium.

If we discard this mechanism, we are left with *conduction* and *radiation.* It turns out that thermal conductivity of stellar matter is too small to be effective in transporting the radiant energy from the interior to the surface. That leaves us with radiation as the only mode of transfer of heat. The idea of *radiative equilibrium,* in which heat is transferred by radiation itself and the temperature distribution is controlled by the flow of radiation, was first invoked by R. A. Sampson in 1894. But a full theory had to wait till the subject of *radiative transfer* was developed by the great German physicist and astronomer, Karl Schwarzschild, in a fundamental paper published in 1906. We shall discuss this in Chap. 3.

Sir Arthur Stanley Eddington
(28 December 1882–22 November 1944)

Eddington was born in Kendal, England, the son of Quaker parents. While at school, he proved to be a brilliant scholar particularly in mathematics and English literature. His performance earned him a scholarship to Owens College, Manchester in 1898. Eddington was greatly influenced by his physics and mathematics teachers, *Arthur Schuster* and *Horace Lamb*. His progress was rapid, winning him several scholarships and he graduated with a B.Sc. in physics with First Class Honours in 1902.

Based on his performance at Owens College, he was awarded a scholarship at Trinity College in the University of Cambridge 1902. Two years later, Eddington became the first ever second-year student to be placed as **Senior Wrangler**—the highest academic distinction for undergraduates. After receiving his B.A. in 1905, he began research on *Thermionic Emission* in the Cavendish Laboratory. This did not go well for him, and he spent time teaching mathematics to first year engineering students, without much satisfaction. But fortunately, this unsatisfactory period was brief!

In 1907, he won the prestigious Fellowship of Trinity College, Cambridge. In December 1912, George Darwin, son of Charles Darwin, died suddenly and Eddington was promoted to his chair as the Plumian Professor of Astronomy and Experimental Philosophy in early 1913. Later that year Eddington was named Director of the Cambridge Observatory. He was elected a Fellow of the Royal Society shortly after.

During World War I, Eddington became embroiled in controversy within the British astronomical and scientific communities. Being a Quaker, Eddington was a pacifist. He struggled to keep wartime bitterness out of astronomy.

He repeatedly called for British scientists to preserve their pre-war friendships with German scientists. Eddington's pacifism caused severe difficulties during the war, especially when he was called up for military service in 1918. He claimed conscientious objector status, a position recognized by the law, but despised by the public. In 1918 the government sought to revoke this deferment, and only the timely intervention of the Astronomer Royal and other high-profile figures kept Eddington out of prison.

Relativity

Eddington was famous for his work in *General Theory of Relativity* which was published by *Einstein* in 1916. During World War I, Eddington was the **Secretary of the Royal Astronomical Society**, which meant he was the first to receive a series of letters and papers from Willem de Sitter regarding Einstein's General Theory of Relativity. Eddington was fortunate in being not only one of the few astronomers with the mathematical skills to understand this theory, but (owing to his international and pacifist views) one of the few at the time who was still interested in pursuing a theory developed by a German physicist! World War I severed many lines of scientific communication and new developments in German science were not well known in England. He quickly became the chief supporter and expositor of Theory of Relativity in Britain. Eddington wrote a number of articles which announced and explained Einstein's theory to the English-speaking world. He also became known for his popular expositions and interpretations of the theory.

He and Astronomer Royal Frank Dyson organized two expeditions to make observations on a solar eclipse in 1919 to conduct the first empirical test of Einstein's theory: *the measurement of the deflection of light by the Sun's gravitational field*. Dyson argued that Eddington's expertise was indispensable for this most important expedition. It was this argument, and the powerful connections that Dyson had, that allowed Eddington to escape prison during the war!

After the war, Eddington travelled to the island of Príncipe near Africa to observe the solar eclipse of 29 May 1919. During the eclipse, he took photographs of the stars in the region around the Sun. According to the theory of General Relativity, stars near the Sun would appear to have been slightly shifted because their light had been curved by Sun's gravitational field. This effect is noticeable only during an eclipse, since otherwise the Sun's brightness obscures the stars.

Eddington's observations, published the next year, confirmed Einstein's theory, and were hailed at the time as a conclusive proof of General Relativity. The news was reported all over the world as a major story. Afterwards,

Eddington embarked on a campaign to popularize relativity and the expedition as landmarks both in scientific development and international scientific relations.

Throughout this period Eddington lectured on relativity, and was particularly well known for his ability to explain the concepts in lay terms as well as scientific. He collected many of these into the publication, *Mathematical Theory of Relativity*, in 1923, which Albert Einstein suggested was 'the finest presentation of the subject in any language'.

Fundamental Theory

During the 1920s, until his death, he increasingly concentrated on what he called the *Fundamental Theory*, which was intended to be a unification of quantum theory, relativity, and gravitation. At first, he progressed along 'traditional' lines, but turned increasingly to an almost numerological analysis of the dimensionless ratios of fundamental constants.

His basic approach was to combine several fundamental constants in order to produce a dimensionless number. In many cases, these would result in numbers close to 1040, its square, or its square root. He was convinced that *the mass of the proton and the charge of the electron were a natural and complete specifications for constructing a Universe* and that their values were not accidental. One of the discoverers of quantum mechanics, Paul Dirac, also pursued this line of investigation, which has come to be known as the Dirac large numbers hypothesis, and some scientists even today believe it has something to it.

Theory of the Stars

Eddington's most important contributions were, however, concerning the nature of the stars. As we shall see in this volume, his insights were deep and his conjectures extraordinarily prescient. But he also made several serious errors of judgment. One of them concerned the fate of massive stars. He rejected the spectacular discovery made by young Chandrasekhar, resulting in a major controversy. We shall discuss this at great length in the next volume of this series. But this most unfortunate controversy did not affect Chandrasekhar's opinion of Eddington. In a lecture delivered at Cambridge University to commemorate Eddington's birth centenary, Subrahmanyan Chandrasekhar described Eddington as the 'most distinguished astronomer of his time'.

Cycling

Eddington was very fond of cycling. He devised a measure of a cyclist's long-distance riding achievements. The Eddington Number in this context is defined as, n, the number of days a cyclist has cycled more than an equal number in distance, namely, n miles. For example an Eddington Number of 30 would imply that a cyclist has cycled more than 30 miles in a day on 30 occasions! Eddington must have been a very good cyclist, and very fond of it. To quote from one of the letters written to S. Chandrasekhar in the evening of his life:

> My n is now 77. It made the last jump a few days ago when I took an 80-mile ride in the fen country. I have not been able to go on a cycling tour since 1940, because it is impossible to rely on obtaining accommodations for the night; so my records advance slowly.

Popular and Philosophical Writings

During the 1920s and 1930s, Eddington gave innumerable lectures, interviews, and radio broadcasts on astronomy, relativity, and later, quantum mechanics. Many of these were gathered into books, such as *Stars and Atoms*, *Nature of the Physical World*, and *New Pathways in Science*. His skillful use of literary allusions and humour helped make these famously difficult subjects quite accessible.

Eddington's books and lectures were immensely popular with the public, not only because of Eddington's clear and entertaining exposition, but also his willingness to discuss the philosophical and religious implications of the new physics. His popular writings made him, quite literally, a household name in Great Britain between the two world wars.

Chapter 3
Eddington's Theory of the Stars

The credit for developing a comprehensive theory of the stars goes to Eddington. He distilled the most significant idea from Lane's work, and built upon it. Although there were very few takers for Lane's idea that the Sun was a *globe of ideal gas*, Eddington was convinced that the matter within all the stars obeyed the perfect gas equation of state, namely, Boyle's law (we shall discuss the validity of this hypothesis in the next chapter). But he rejected Lane's notion that the outward transport of heat is due to convection. Instead, he introduced two basic elements into the theory of the stars. These are:

1. Radiation pressure must play an increasingly important role in maintaining the equilibrium of stars of increasing mass.
2. Stars must be in radiative equilibrium. The outward flow of radiation is the main mode of heat transport.

These two insights had important bearing on the structure of stars. Before examining some of the predictions of the theory, let us be clear about the two notions mentioned above.

Radiation Pressure

You will recall that radiation has momentum E/c, where E is the energy and c is the velocity of light (in the quantum picture, the momentum of a photon is $h\nu/c$, where h stands for Planck's constant and ν is the frequency of the photon). Since momentum is associated with radiation, it must exert pressure just as gas particles do. Let us consider a special kind of radiation, popularly known as *black body radiation*. This is just radiation in an enclosure with absorbing walls maintained at a temperature T. We have already encountered the very special properties of this radiation while discussing Kirchoff's laws. Given enough time, the radiation in the cavity will come to thermal equilibrium with the walls. It will be isotropic as regards the direction of flow and will be characterized uniquely by the temperature of the walls of the cavity.

G. Srinivasan, *What are the Stars?* Undergraduate Lecture Notes in Physics,
DOI: 10.1007/978-3-642-45302-1_3, © Springer-Verlag Berlin Heidelberg 2014

An important result from the nineteenth century is that *the energy density of radiation in the cavity is proportional to the fourth power of the absolute temperature,*

$$E = aT^4 \tag{3.1}$$

where a is a universal constant known as *Stefan's constant*. The above relation is known as *Stefan's law*. The pressure exerted by this radiation is

$$\boxed{p_R = \frac{1}{3}aT^4} \tag{3.2}$$

You may be puzzled by why the pressure of radiation is *one-third* of the energy density and not two-thirds, as we had remarked earlier while deriving Boyle's law. There are many ways of understanding this, but I shall not digress into that now. Briefly, this is the reason for the difference. As you know, in 1905, Einstein taught us that light behaves both as waves, as well as particles. We now call these corpuscles of light *photons*. The important thing is that these photons have *zero rest mass*. Whenever we are dealing with an enclosure containing particles with zero rest mass (such as photons, neutrinos, and the like), the pressure in the enclosure is *one-third* of the energy density. If one is dealing with a gas of massive particles then the pressure is *two-thirds* of the energy density.

Returning to radiation inside a star, we can assume that the radiation has come to thermal equilibrium with matter and, therefore, the above two relations apply to it. The star may not be covered with opaque walls, but the interior of the star is opaque to the kind of radiation existing there. Therefore one may safely assume that the radiation inside will have all the characteristics of true thermal radiation.

Clearly, the main thrust of this chapter is going to be the interaction of radiation with matter; how the radiation generated near the centre diffuses out. Therefore, before proceeding further with Eddington's ideas let us pause to define some important concepts.

Scattering cross-section, opacity and mean free path

Before proceeding further, let us familiarise ourselves with these three important concepts.

Imagine a flux of particles, or radiation, attempting to pass through a target. And let there be *n obstacles per unit volume* in the target. These could be atoms, molecules, dust particles, and so on, which could absorb or scatter the particles in the incident beam. Let σ be the *effective cross-sectional area for collision* of each one of these scattering centres (usually referred to simply as *scattering cross-section*). The unit of σ is cm^2. It is important to appreciate that this is not necessarily the actual projected physical area of the atom,

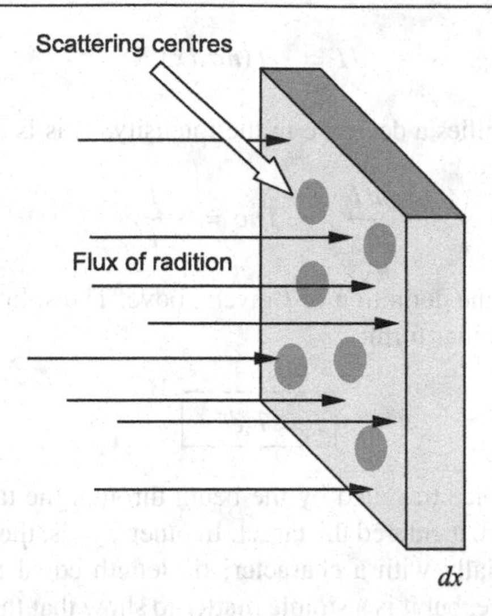

molecule and so on; it could be much larger than that. Imagine, for example, an electron incident on an ion. The electron will be deflected by the electric field associated with the charge on the ion. Therefore, the radius of the *effective sphere of influence* of the ion would be much larger than the size of the ion.

The mean free path l and the scattering cross-section σ

Using the characteristics of the target, namely, there are n scatterers per unit volume, each with an effective area equal to σ, let us construct a quantity with the dimension of a *length*:

$$l = \frac{1}{n\sigma}.$$

This length l is defined as the mean free path.

To see the significance of this, let us consider a slab of the target of area $L \times L$ and thickness dx Its volume would be $L^2 dx$. The typical number of absorbers in this slab would be the number of absorbers per unit volume, n, multiplied by the volume, that is, $nL^2 dx$. *The probability that a beam particle will be stopped in that slab is the net area of the stopping atoms divided by the total area of the slab.*

P (absorption within dx) =

$$\frac{\text{Total Area of the Absorbers}}{\text{Area of the Slab}} = \frac{\sigma n L^2 dx}{L^2} = n\sigma dx.$$

The decrease in beam intensity as it goes through the slab of thickness dx equals the incoming beam intensity multiplied by the probability of being stopped within the slab

$$dI = -I(n\sigma dx).$$

The minus sign signifies a decrease in the intensity. This is an ordinary differential equation

$$\frac{dI}{dx} = -In\sigma = -\frac{I}{l}.$$

Here we have used the definition of l given above. The solution of this simple equation has the familiar form:

$$\boxed{I = I_0 e^{-\frac{x}{l}}}$$

where x is the distance travelled by the beam through the target and I_0 is the beam intensity before it entered the target. In other words, the incident intensity decreases exponentially with a characteristic length equal to l. We shall not pause to prove it here, but it is a simple matter to show that this length, l, is also the *mean distance* travelled by the photon/particle before it is stopped.

Mass absorption coefficient κ

Sometimes, instead of the scattering cross-section, σ, one introduces the *absorption coefficient per unit mass*, κ. Astronomers refer to it as the *opacity* of the material. The units of κ are cm^2 g^{-1}. Earlier we had written the decrease in the beam intensity as $dI = -I(n\sigma dx)$. In terms of the opacity this can be written as

$$dI = -I(\kappa\rho dx).$$

Remember that the mass of the target slab of unit area and thickness dx is simply the volume multiplied by the density, ρdx. Therefore the absorption coefficient of the slab is $\kappa\rho dx$. Clearly, $n\sigma = \kappa\rho$. Therefore, the mean free path can be defined either in terms of the cross-section σ or the mass absorption coefficient, κ:

$$l = \frac{1}{n\sigma} = \frac{1}{\kappa\rho}.$$

As we shall soon see, the opacity of matter will be a function of *frequency*. Therefore, whenever we simply write κ, it must represent an *average over the frequency* of some sort.

Radiative Equilibrium

Let us now derive the condition for radiative equilibrium to be established. Pick any radial direction in the star, and call it the x-axis, and let the positive direction of this axis be along the temperature gradient. Consider a slab of stellar material of thickness dx and area equal to one square centimetre held normal to the x-axis (see Fig. 3.1).

Let the temperature of the two faces of the slab be T and $T + dT$, respectively. Since pressure is force per unit area, the force exerted by radiation on the two faces is $+p_R$ and $-(p_R + dp_R)$. The resultant force in the direction of the temperature gradient is $-dp_R$. We have adopted the convention that the force is positive if it is in the direction of gravity and negative if directed outwards.

This resultant force imparts momentum to the slab. For the slab to be in equilibrium, it must utilize this momentum in some fashion; otherwise, the slab will be set in motion. What the material of the slab does is to *absorb* this momentum and *use it to supplement the gas pressure* in its attempt to support itself against gravity.

Next we have to calculate the x-component of the momentum absorbed by the material in the slab. Let us first introduce the mass absorption coefficient κ. This is the coefficient of absorption *per gram* of matter. Let F be the flux of radiation incident on the slab (measured in *ergs per square centimetre per second*). The fraction of the flux absorbed by the slab will be $F\kappa\rho dx$, where ρ is the density of matter in the slab. Since the area of the slab is unity, and its thickness is dx, the mass of the slab is just ρdx (see Fig. 3.2). The x-component of the momentum absorbed by the material per unit time is

$$F\kappa\rho dx/c \tag{3.3}$$

where c is the velocity of light. (Interestingly, the above result holds even if the radiant flux is incident obliquely. If the angle of incidence is θ then the distance travelled through the slab is increased to $dx \sec\theta$. So the energy absorbed in the slab increases by $\sec\theta$. But the x-component of the momentum absorbed remains the same as above because we have to multiply by $\cos\theta$ to obtain the x-component.)

Finally, we want to calculate the net momentum absorbed by the slab per unit time. Remember that radiation is incident on the slab from both sides. Let us denote the flux from the left (outward flowing) by F_+ and the flux from the right by F_-. The *net outward flux* is given by

$$F = F_+ - F_-, \tag{3.4}$$

and the net positive momentum gained by the slab is $F\kappa\rho dx/c$. Earlier we said that for the slab to be in radiative equilibrium, the momentum gained per second by the slab must be fully absorbed by the matter contained in it. Hence,

$$-dp_R = F\kappa\rho dx/c \tag{3.5}$$

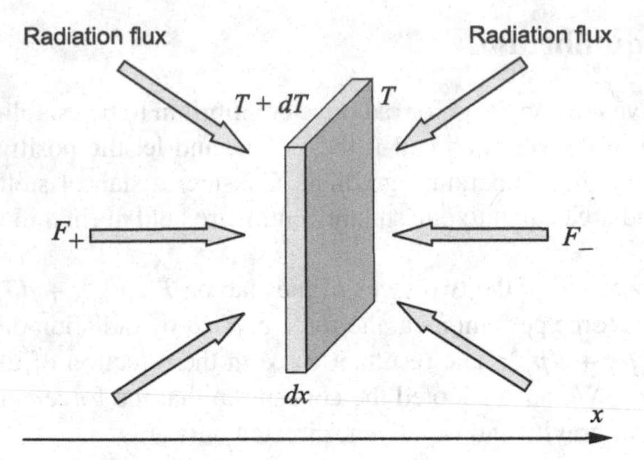

Fig. 3.1 Consider a slab of stellar material of unit area and thickness dx along some radial direction which we shall call x. Radiation passes through this slab from both sides. Let the temperature on the two faces be $T + dT$ and T, respectively. Consequently, the flux of radiation and the pressure of radiation on one side will be more than on the other. The resultant pressure (or force, since the slab has unit cross-sectional area) in the direction of the temperature gradient would be $-dp_R$. By convention, we take the force to be positive in the direction of gravity, and hence the resultant force due to radiation has a negative sign. This resultant force due to radiation will impart momentum to the slab, which it must utilize in some way so that it can be in *radiative equilibrium*. Otherwise the slab will start moving in the x direction

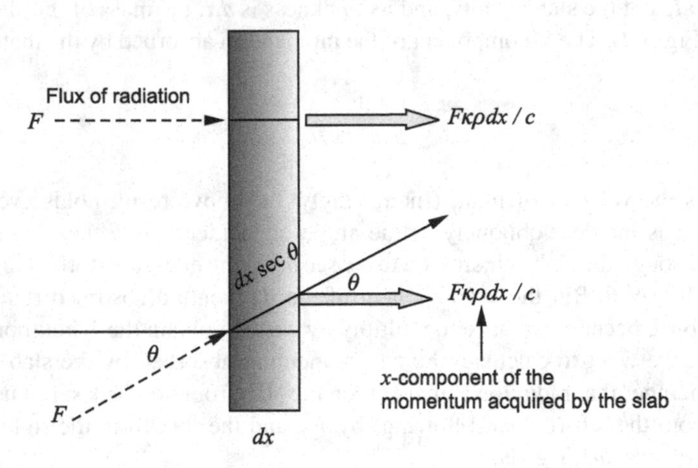

Fig. 3.2 The fraction of the radiation flux F absorbed by the slab will be equal to the flux multiplied by the mass absorption coefficient *per unit mass* multiplied by the *mass* of the slab, that is, $F\kappa\rho dx$. Therefore the x-component of the momentum acquired by the slab per unit time will be $F\kappa\rho dx/c$. Incidentally, this remains the same regardless of whether the radiation is incident normally or at an angle. For the slab to be in equilibrium,this x-component of the momentum absorbed from the radiation must be equal to $-dp_R$. The slab uses this momentum absorbed from the radiation to *supplement the gas pressure* in supporting itself against gravity. This is the principle of *radiative equilibrium*

or

$$F = -\frac{c}{\kappa\rho}\frac{dp_R}{dr}.$$

(We have replaced x by the radial co-ordinate r). Substituting for the radiation pressure from Stefan's law, $p_R = \frac{1}{3}aT^4$, we get for the net *outward flux*:

$$F = -\frac{ac}{3\kappa\rho}\frac{dT^4}{dr}. \tag{3.6}$$

This is the famous result obtained first by Eddington. It says that *net flux of radiation* is directly proportional to the *pressure gradient* and inversely proportional to the *opacity* of the stellar matter (Eddington called $\kappa\rho$ the *obstructive power of the material screen* through which the radiation is forced).

Basic Equations of Stellar Structure

Let us now gather together the various equations that constitute a theory of the stars.

1. Hydrostatic equilibrium

As we have derived in Chap. 2, *Stars as Globes of Gas,* for a nonrotating star the assumption of hydrostatic equilibrium gives:

$$\frac{dP}{dr} = -\frac{GM(r)\rho}{r^2} \tag{3.7}$$

where P is the *sum of gas pressure and radiation pressure,*

$$P = p_g + p_R, \tag{3.8}$$
$$p_g = nk_BT = \frac{\rho k_B T}{\mu m_H},$$
$$p_R = \frac{1}{3}aT^4.$$

In the above equations,

n = number density of particles
ρ = mass density
μ = mean molecular weight
m_H = mass of the hydrogen atom ($= 1.67 \times 10^{-24}$ g)
k_B = Boltzman's constant
a = Stefan's constant.

You will notice that in the equation for hydrostatic equilibrium the pressure that balances gravity is the *total* pressure. This is as it should be if the star is in radiative equilibrium. We said that the outward-directed momentum of the radiation is absorbed by the stellar material. The star is now using this to supplement gas pressure in its attempt to support itself against gravity.

Since the mass contained in any spherical shell of radius r and thickness dr is $dM(r) = 4\pi r^2 dr\rho$, the rate of change of $M(r)$ with r is

$$\boxed{\frac{dM(r)}{dr} = 4\pi r^2 \rho.} \tag{3.9}$$

This is known as the *mass equation.*

2. Thermal equilibrium

In a similar manner, we need an equation describing how the energy radiated by the surface is compensated by the energy generated in the interior. Let ε be the rate of energy generation *per gram per second.* This will clearly depend upon the temperature, density and chemical composition. The total luminosity, $L(r)$, crossing an imaginary surface of radius r is given by

$$L(r) = \int_0^r \varepsilon\rho 4\pi r^2 dr \tag{3.10}$$

Therefore the change of $L(r)$ with r is given by

$$\boxed{\frac{dL(r)}{dr} = \varepsilon\rho 4\pi r^2.} \tag{3.11}$$

3. Radiative equilibrium

While discussing the idea of radiative equilibrium, we had earlier derived the expression for the net radiative flux crossing a unit area at a distance r from the centre of the star (erg per square centimetre per second). To get the expression for the luminosity crossing an imaginary surface of radius, r, we merely have to multiply this by $4\pi r^2$, which is the surface area of this sphere. Therefore,

$$\boxed{L(r) = -\frac{ac}{3\kappa\rho}\frac{dT^4}{dr}4\pi r^2.} \tag{3.12}$$

Let us gather together all the four equations of stellar structure:

$$\frac{dP}{dr} = -\frac{GM(r)\rho}{r^2},$$

$$\frac{dM(r)}{dr} = 4\pi r^2 \rho,$$

$$\frac{dL(r)}{dr} = \varepsilon \rho 4\pi r^2,$$

$$L(r) = -\frac{ac}{3\kappa\rho}\frac{dT^4}{dr}4\pi r^2.$$

These four equations constitute Eddington's theory of the stars.

Solution of the Equations of Stellar Structure

To derive the expressions for temperature and the density distribution which is capable of maintaining itself automatically, notwithstanding the continual transfer of heat from one part to the other, we have to solve the above equations self-consistently. The four basic inputs needed to solve these are:

1. Equation of state
2. The chemical composition
3. Opacity
4. The rate of energy production

1. Equation of state

This gives the pressure as a function of density and temperature $P = P(\rho, T)$. In Eddington's theory, this is the sum of gas pressure and radiation pressure. Following Lane, Eddington assumed that the gas can be well described by the *perfect gas law* $p_g = nk_B T = \rho k_B T/\mu m_H$, and that the radiation is black body radiation, whose pressure is given by *Stefan's law*, $p_R = \frac{1}{3}aT^4$.

2. Chemical composition

So far we have remained silent on the chemical composition of the stars, but we need to address this now. After all, as Eddington put it: 'An architect before pronouncing an opinion on the plans of a building will want to know whether the material shown in the plans is to be wood or steel or tin or paper'. It is true that we can find out a great deal about the chemical composition of the Sun's atmosphere by studying the solar spectrum, but we know very little of the interior. In any case, when Eddington was developing his theory there was no way of estimating the relative abundance of the elements. The prevailing opinion at the time was that the Sun's composition must be similar to the Earth's, with the predominance of heavy elements like iron.

Table 3.1 The most abundant elements in the Sun

Element	Relative mass (%)
Hydrogen	70
Helium	28
Carbon	0.41
Nitrogen	0.10
Oxygen	0.91
Neon	0.14
Magnesium	0.06
Silicon	0.06
Sulphur	0.04

A proper theoretical framework to estimate the abundance of elements had to wait for the theory of ionization of atoms due to M. N. Saha.

Soon after Saha published his theory, Cecelia Payne, one of the very few women astronomers, carefully analysed stellar spectra and came to the remarkable conclusion that the Sun and stars are composed almost entirely of gaseous hydrogen. This radical conclusion by a young student was not readily accepted by the high priests of astronomy, notably Henry Norris Russell of Princeton University, one of the most distinguished astronomers of that era. But some instinct must have told Russell that they had all got it wrong, for he reanalysed the data, using Saha's theory, and concluded that Cecelia Payne was right after all!

The modern picture is that hydrogen accounts for 70 % of the mass of the Sun. The element helium, which you will recall was first discovered in the Sun, accounts for 28 %. The rest of the elements taken together contribute only about 2 % to the mass of the Sun (see Table 3.1).

Let us return to Eddington. He did not wait for the above story to unfold. He proceeded with his theory by assuming that the Sun was mainly composed of heavy elements, like iron. But he did not care for the detailed abundance of the heavy elements. And the reason was the following. As we have already discussed, Eddington was convinced the pressure inside the Sun was enormous and that the temperature was approximately 10 million K. Under these circumstances the atoms would be severely mutilated. Remember that by this time Bohr has elucidated the structure of the atoms. In his picture, an atom resembled the solar system; it consisted of tiny positively charged nuclei (of size approximately 10^{-13} cm) around which negatively charged electrons revolved in discrete orbits whose radii were approximately 10^{-8} cm. In neutral atoms, the number of satellite electrons is equal to the atomic number Z of the element. At high pressure and high temperature a phenomenon known as *ionization* occurs by which the satellite electrons are successively knocked off the atom. This happens for two reasons.

At a temperature of 10 million K, the heat radiation will be mainly *soft* X-rays with wavelengths between 3 and 9 Å. Let us see why this is so. You may remember that the spectrum of radiation from a black body is sharply peaked at a certain wavelength, which depends upon the temperature of the black body (see Fig. 3.3). The wavelength

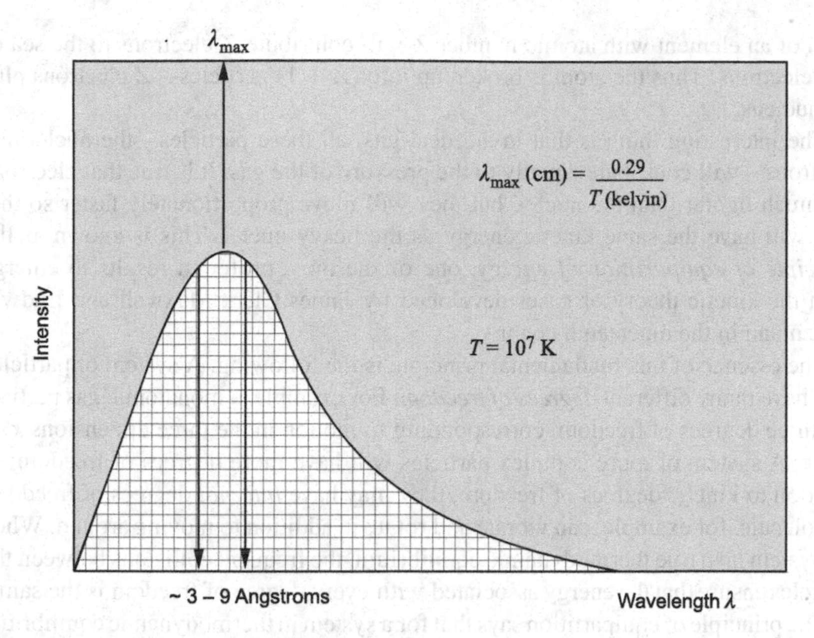

Fig. 3.3 The spectrum of radiation from a *black body* is quite sharply peaked at a wavelength which is inversely proportional to the temperature of the body. This is known as *Wien's displacement law.* For a body at 10 million k, most of the radiated energy is in the soft X-ray region of the spectrum. The present temperature of our universe is approximately 3 k. Hence the thermal radiation that fills the universe peaks at a wavelength of 1 mm

λ_{max} at which the intensity is maximum is related to the temperature by the relation

$$\lambda_{max}(\text{cm}) = \frac{0.29}{T(K)}. \tag{3.13}$$

This is known as *Wien's Displacement Law.*

Verify that for a temperature of 10 million K, the maximum occurs at a wavelength of approximately 3 Å. These soft X-ray photons have enough energy to knock out the electrons from their orbits; this is just the *photoelectric effect* that Einstein had explained. In lighter elements, the energy needed to eject the electrons is sufficiently small, so that almost all the electrons will be knocked out. In heavy elements, the innermost electrons are so tightly bound to the nucleus that they will survive.

The second channel for ionization is this. The electrons knocked out the atoms by the X-ray photons wander around as free particles. They will have kinetic energy equal to the difference between the energy of the ionizing photon and the energy it has spent in knocking out the electron from the atom. These wandering energetic electrons can also, provided they have enough energy, dislodge the bound electrons from their orbits; this is known as *collisional ionization.*

Deep in the interior of a star, one can expect complete ionization of the atoms and a great simplification occurs—the atoms will be stripped off all their electrons. An atom of an element with atomic number Z will contribute Z electrons to the sea of free electrons. Thus the atom is broken up into $(Z + 1)$ particles—Z electrons plus the nucleus.

The interesting thing is that in an ideal gas, all these particles—the nuclei and electrons—will contribute equally to the pressure of the gas. It is true that electrons are much lighter than the nuclei, but they will move proportionately faster so that they will have the same kinetic energy as the heavy nuclei. This is known as the *principle of equipartition of energy*, one of the most profound results to emerge from the kinetic theory of gases developed by James Clerk Maxwell and Ludwig Boltzmann in the nineteenth century.

The essence of this fundamental principle is the following. A system of particles may have many different *degrees of freedom*. For example, a monatomic gas particle has three degrees of freedom, corresponding to motion in the three dimensions x, y and z. A system of more complex particles will have more degrees of freedom; in addition to kinetic degrees of freedom, they may have *internal* degrees of freedom. A molecule, for example, can vibrate and rotate in addition to moving around. When the system is in true thermodynamic equilibrium, the frequent collisions between the particles ensure that the energy associated with every degree of freedom is the same.

The principle of equipartition says that for a system in thermodynamic equilibrium at a temperature T, *the energy per degree of freedom is* $\frac{1}{2}k_B T$. If there are three translational degrees of freedom then the kinetic energy per particle is $\frac{3}{2}k_B T$. The following point must, however, be borne in mind. The energy of any given particle, of course, constantly changes because of collisions. Therefore, the *equipartition of energy* refers to the *average* energy of any particle, with the average being taken over a long enough period of time. Clearly, *the statement that the average energy of the particles is* $\frac{3}{2}k_B T$ *is irrespective of the mass of the particle*. The speed of the particles will adjust such that

$$\frac{3}{2}k_B T = \frac{1}{2}mv^2. \tag{3.14}$$

This is why in an ideal gas all the particles—nuclei and electrons—will contribute equally to the pressure of the gas. And that is the reason why in Boyle's law the pressure is expressed in terms of the *number density* of particles. In the present context, however, it is more appropriate to express the pressure in terms of the *mass density* ρ. If our gas consisted of particles of only one type then

$$\text{number density} = \frac{\text{mass density}}{\text{mass of the particle}}.$$

The stellar plasma, however, consists of electrons and nuclei of different elements. Therefore the correct things to do would be to divide the mass density by the average mass of the independent particles:

$$\text{number density} = \frac{\text{mass density}}{\text{average mass of the particles}}.$$

It is to define this average mass that one needs to know the chemical composition of the plasma. It is customary to introduce the notion of the *mean molecular weight*, μ, in defining the relation between the *number density* of independent particles, n, and the *mass density*, ρ,

$$n = \frac{\rho}{\mu m_H}, \tag{3.15}$$

where m_H is the mass of the proton. The terminology *molecular weight* is borrowed from chemistry and is a misnomer here. In the present context, the term *molecule* really refers to the independent particles of our gas, nuclei of different species and the electrons. It should be clear from the above equation that μm_H is defined as *the average mass of the independent particles of the gas.* Remember that our gas consists of electrons and nuclei of different species. Therefore, the average mass per particle will depend upon the chemical composition or the relative abundance of the elements. But these are matters of detail. Eddington argued that the detailed composition really did not matter as long as the heavy elements predominate; like most astronomers at that time, he believed that the composition of the stars must be similar to what we find on Earth. He argued that in that case μ will be approximately equal to 2. His line of reasoning may be understood as follows.

In general, an atom of an element with atomic mass number A and atomic number Z will contribute $(Z + 1)$ particles; Z electrons and 1 nucleus. Therefore the average mass per particle would be

$$\mu m_H = \frac{A}{(Z+1)} m_H. \tag{3.16}$$

If the atomic number Z is much greater than 1, which is the case for heavy elements, then $(Z + 1)$ in the denominator can be replaced by Z. It is a well known fact that, barring hydrogen, A/Z for most elements is approximately 2, so that μ is approximately equal to 2, as long as hydrogen is *not* the predominant constituent (as was believed at that time). This is why Eddington did not bother to ascertain the detailed chemical composition of the star and simply assumed that $\mu = 2$. If the star consists essentially of hydrogen then $\mu = 1/2$; in the above formula, substitute $A = 1$ and $Z = 1$.

To summarize this section, Eddington assumed that the composition was dominated by the heavy elements and, therefore, the mean molecular weight, μ, was approximately equal to 2. He was wrong, of course! We now know that hydrogen and helium account for 98 % of the composition.

3. Opacity of stellar matter

Let us now turn to the opacity, or the obstructive power, of stellar matter. Before discussing the physical mechanism that hinders the outflowing radiation, let us try to understand the nature of the matter inside a star. We know that hydrogen and helium atoms will be fully ionized, but the heavier atoms will still retain their innermost electrons since these are very tightly bound to the nucleus; the more loosely bound outer electrons will be knocked out without any difficulty. The electrons liberated from the atoms will be rushing around at fantastic speeds. So we have a plasma consisting of electrons, bare nuclei, and heavy ions. This *plasma* coexists with radiation, which is mostly in the form of X-rays. There is no better way to visualize what goes on inside a star than listening to Eddington's marvellous description:

> We can now form some kind of a picture of the inside of a star—a hurly burly of atoms, electrons and ether-waves. Dishevelled atoms tear along at 100 miles a second, their normal array of electrons being torn from them in the scrimmage. The lost electrons are speeding a hundred times faster to find new resting places. Let us follow the progress of one of them. There is almost a collision as an electron approaches an atomic nucleus, but putting on speed it sweeps round in a sharp curve.
> Sometimes there is a side-slip at the curve, but the electron goes on with increased or reduced energy. After a thousand narrow shaves, all happening within a thousand millionth of a second, the hectic career is ended by a worse side-slip than usual. The electron is fairly caught, and attached to an atom. But scarcely has it taken up its place when an X-ray bursts into the atom. Sucking up the energy of the ray, the electron darts off again on its next adventure.
> 'And what is the result of all this bustle? Very little. The atoms and the electrons for all their hurry never get anywhere; they only change places. The ether-waves are the only part of the population which accomplish anything permanent. Although apparently darting in all directions indiscriminately, they do on the average make a slow progress outwards ... But slowly the encaged ether-waves leak outwards as through a sieve ....

Let us recall that one of the main objectives of any theory of the stars is to calculate the luminosity of the star. Therefore, it is this leakage of radiation that concerns us. The radiation would like to flow out, urged by the temperature gradient, but it is hindered by the process of absorption and emission during its encounter with the ionized atoms and electrons. It is this process that couples matter and radiation. If it were not for this the radiation will stream out. It is this repeated absorption and emission that forces the radiation to do a *random walk* and diffuse out very slowly (in tens of thousands to millions of years, depending on the mass of the star). We have already seen that for a star in radiative equilibrium the luminosity depends not only on the temperature gradient, but also inversely on the *opacity*.

We now turn to a discussion of the opacity of stellar matter, which consists of fully or partially ionized atoms and electrons. And the radiation is mainly soft X-rays (remember that for a black body at a temperature of approximately 10^7 K, the spectrum peaks in the soft X-ray region). This is a complicated subject, and we shall not get into the details. Instead we shall make some general remarks that should enable you to get a feeling for the underlying physics. The absorption of X-rays by atoms has been well studied in the laboratory. The main process is the ionization of atoms; X-rays of sufficient energy can be absorbed by the atom to eject one or more of its orbiting electrons. Although the same mechanism will be operative in the stars, on very general grounds one can anticipate that the opacity of stellar matter would be *less* than that of terrestrial matter. There are two reasons for this.

First, stellar atoms are badly mutilated. We expect hydrogen and helium to be fully ionized, and the heavy elements stripped of all but the ten innermost electrons (two in the K shell and eight in the L shell). This reduces the absorbing power of stellar atoms compared to terrestrial atoms which are essentially neutral and, therefore, have their full complement of electrons.

There is another reason and let us try to understand this. The point to bear in mind that the very act of absorption makes the atom ineffective until the atom is *repaired*. To repair the absorbing mechanism the atom must capture one of the wandering electrons, in order to replace the lost electron. This is true in the laboratory also. But the laboratory X-ray source is quite weak and there is plenty of time for the atom to repair itself before the next X-ray photon invades it, but in a star, the intensity of X-rays is enormous. As Eddington put it, 'It is like an army of mice marching through your larder springing the mouse traps as fast as you can set them. Here it is the time wasted in resetting the traps—by capturing electrons—which counts, and the amount of the catch depends almost entirely on this'.

This is the reason stellar opacity depends on the density, whereas terrestrial opacity does not. Inside a star, as the density increases, the chance of an atom capturing a free electron increases, thus decreasing the time needed to repair the traps.

After this qualitative discussion of opacity, let us be a little more systematic and list the various mechanisms that contribute to the absorption coefficient. You may skip this discussion if you have not studied some atomic physics.

There are four processes that contribute to the opacity.

1. Photoionization or photoelectric effect
2. Bound–bound transitions
3. Free–free transitions
4. Electron scattering

The relative importance of these depends upon the prevailing condition. Let us now briefly discuss each one of these.

A. Photoionization

The theory of photoionization or photoelectric absorption was worked out by the great Dutch physicist H. A. Kramers, in 1923 (in astronomical literature, this is referred to as *bound–free transition*). Characteristically, Eddington was quick to realize that this will be the principal physical process contributing to stellar opacity. The incident photon knocks out one or more electrons from a neutral atom or an ion (see Fig. 3.4). Neutral atoms are unlikely to exist in the million degree stellar plasma. The light elements like hydrogen and helium would be fully ionized. But the heavy elements would only be partially ionized. Eddington realized that the ionization of the innermost electronic shells (K and L shells) would contribute to opacity. Although the process is rather simple, to calculate the net absorption coefficient per gram of matter is a rather complicated matter.

Consider this. For simplicity, let us first concentrate on atoms of one particular element. Nevertheless, it will be a mixture of atoms with varying degree of ionization (that is, different number of electrons knocked out of them). In a particular ion of that element, the X-ray photon may be absorbed by an electron in any one of the various energy levels (or orbits). To add to this, the absorption depends upon the frequency of radiation. Consider an electron whose principal quantum number is n. The absorption coefficient for *one electron in one atom* is approximately given by:

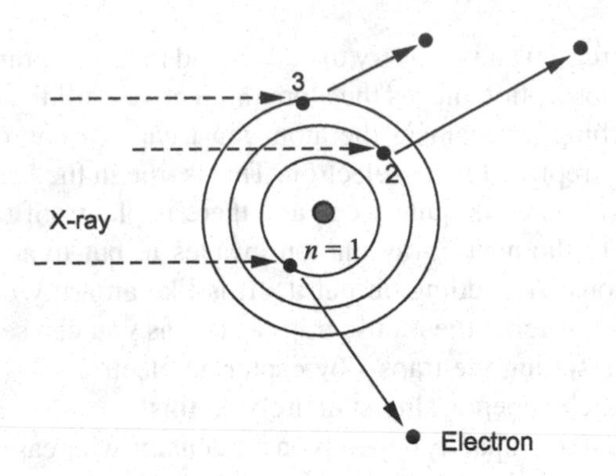

Fig. 3.4 Photoionization is a very important process contributing to the absorption of energy from radiation. While the atoms of light elements like hydrogen and helium will be fully stripped of their electrons, the atoms of heavy elements will still retain the more tightly bound electrons in the inner orbits. The incident X-rays can knock out one or more of these electrons from the occupied levels, provided they have energy which is greater than the binding energy of that particular level. Therefore, the absorption coefficient from any given level will be zero for energy less than the binding energy. Beyond this threshold, the absorption decreases as the cube of the frequency of the photon (this is sketched in Fig. 3.5). The electrons liberated from the atoms may collide with other atoms and can, in principle, cause ionization provided they have sufficient kinetic energy

$$a_{bf} = (\ldots)\frac{1}{n^5}\frac{1}{\nu^3} \tag{3.17}$$

where ν is the frequency of the radiation. In the above formula, we have suppressed the details for the sake of bringing out just the salient features. Consider the cross-section for knocking out an electron from level n. The important feature to notice is that the absorption coefficient decreases inversely as the cube of the frequency. Obviously, this formula can only hold for frequencies for which the corresponding photon energies exceed the ionization energy χ_n of this orbit, for only then can it knock out an electron from this orbit. In other words, the cross-section is finite only for frequencies greater than a critical value, which, in turn, depends on n.

$$h\nu > \chi_n = (\ldots.)\frac{1}{n^2} \tag{3.18}$$

You will remember from your introductory course on atomic physics that this is just the expression for the binding energy of the various levels of a hydrogen atom. We can get away with using this because even the heavy elements will be almost completely ionized, and the ion will look very much like a *hydrogen atom*, but with an effective charge which is different; this is known as the *hydrogenic approximation*. Hence, for a specific bound–free transition, *the absorption coefficient is zero at low frequencies, jumps to its maximum value at the critical frequency, and decreases inversely as the cube of the frequency.* For a specific ion, as a function of frequency the absorption

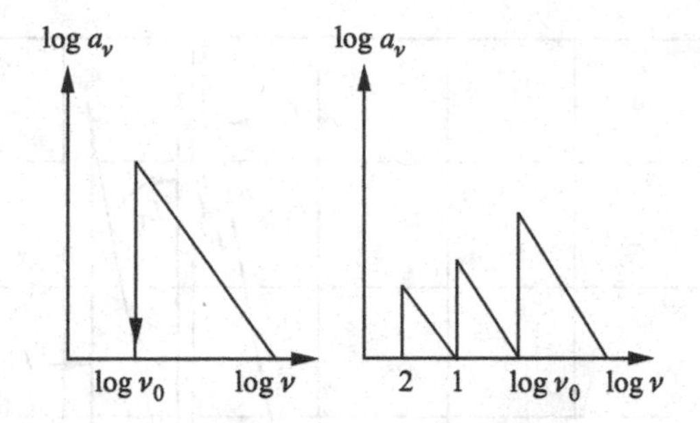

Fig. 3.5 The absorption coefficient due to photoionization plotted as a function of the frequency of the radiation. Shown on the *left* is the absorption coefficient due to one electron in an energy level with binding energy equal to $h\nu_0$. The absorption sharply sets in at the threshold frequency υ_0 and then decreases as ν^{-3}. Notice that this is a log–log plot. That is why it looks like a *sawtooth*. If there are many occupied levels, then the absorption coefficient will have one sawtooth feature for every level. This is shown on the *right-hand-side* graph

coefficient will have several such *sawtooth* features, each corresponding to absorption by electrons in different energy levels of the ion (see Fig. 3.5).

In Fig. 3.6 we have shown the experimental data for lead. Please note that in this particular figure, the absorption coefficient is plotted as a function of wavelength, and not frequency. Also, it is a linear plot and not a log–log plot; this is why it looks different from what is sketched in Fig. 3.5.

Now let us return to the element in question. The story is far from complete. So far we have considered only one species of ion. There will be other species of ions with different degrees of ionization, depending on the density and temperature (that is, different number of electrons knocked out). Corresponding to every stage of ionization there will be a series of *sawtooth* continuum (like those shown in Fig. 3.5). If you think this is a mess, remember that all this is just for one element! The absorption coefficient of a mixture of heavy elements will naturally be quite complicated as a function of frequency, as well as with temperature and density.

B. Bound–bound transitions

There is another absorption mechanism involving ions. Here, an electron in one of the occupied levels absorbs the photon and jumps to one of the higher empty levels (see Fig. 3.7). You might object to this being a genuine absorption mechanism. After all, the electron can, and will, jump back to the lower level emitting a photon. So the photon is not lost. The point is that the re-emitted photon will not necessarily be moving in the same direction as the incident beam. In that sense, there is a decrease in the beam intensity; there has been absorption. But please remember that this mechanism is important only at certain frequencies; the frequency of the photon must match the energy level difference between the final state of the electron and its initial state. If the energy levels of the atom are sharp, then this is a serious restriction. However, because of collisions between the atoms, the energy levels will be *significantly broadened*. Because of this, this mechanism can be important at

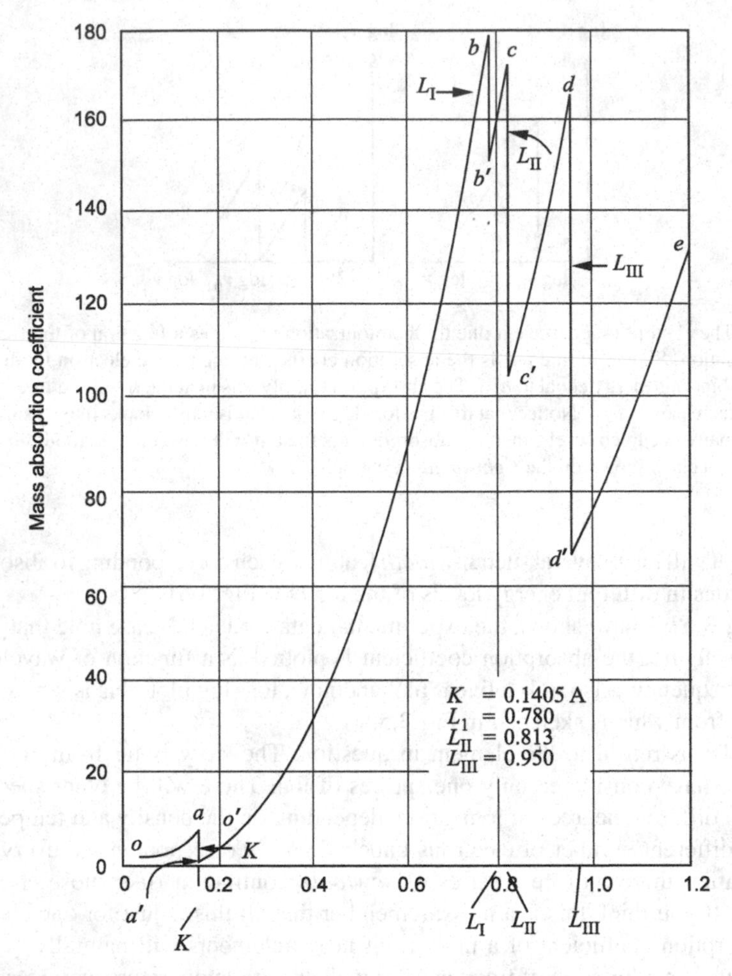

Fig. 3.6 This experimental data on the X-ray absorption edges in lead was obtained by F. K. Richtmyer in 1925, and is reproduced from his classic book, *Introduction to Modern Physics*, published by McGraw–Hill Book Company in 1928. Two differences compared to Fig. 3.5 should be noted. On the y-axis, the absorption coefficient is plotted on a linear scale. Further, plotted on the x-axis is the *wavelength* and not the frequency. The *absorption edges* due to ionisation of electrons in the K and L shells are clearly seen. The K shell corresponds to $n = 1$, and there are no subshells. The L shell corresponds to $n = 2$, and there are three subshells associated with this

times. For example, in the outer layers of the star where the temperature can be less than 10^6 K, this mechanism can make a significant contribution to the total opacity.

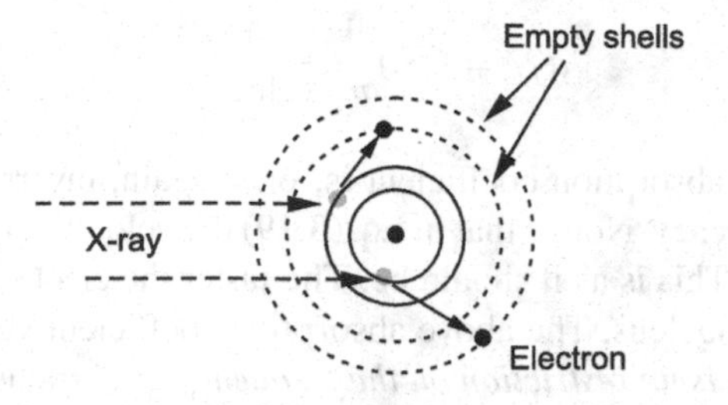

Fig. 3.7 In partially ionized atoms, there is another mechanism that contributes to absorption. The X-ray photon can kick an electron from an occupied to an unoccupied level. Clearly, the absorption coefficient will be finite only at some discrete frequencies which correspond to the *difference in energy between the final state of the electron and its initial state.* This makes this a less-important mechanism than photoionization, discussed in Fig. 3.6. But this process can be important in the outer regions of the star where there is significant ultraviolet flux (at temperatures less than a million degrees, the black body spectrum will peak in the ultraviolet). While the UV photons will not have enough energy to knock an electron from the ion, it can cause the bound–bound transitions shown here

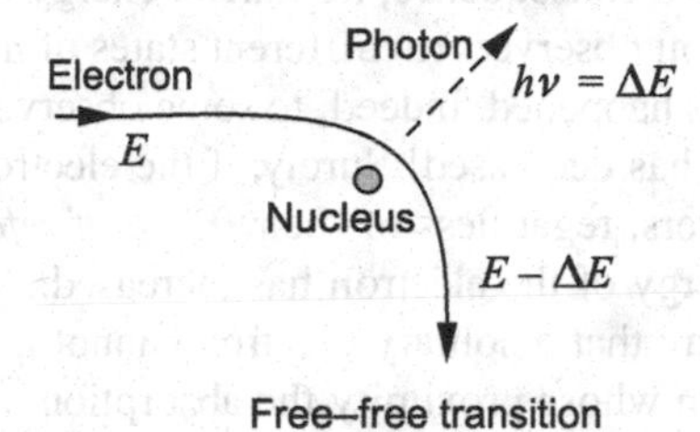

Fig. 3.8 Although a free electron can neither emit nor absorb a photon, an electron in the vicinity of an atomic nucleus can do so. This sketch depicts the *emission* of a photon when the electron experiences *acceleration* while it sweeps past the nucleus. This process is usually referred to as *Bremsstrahlung* or *brake radiation.* This is the mechanism by which the X-ray continuum in the star, for example, is produced. The inverse of this would be the absorption of a photon

C. Free–free transitions

In addition to the above two mechanisms, we have to consider what are known as free–free transitions. Here, while sweeping past a nucleus, an electron absorbs a photon and changes its kinetic energy. In quantum mechanical jargon, it goes from one free state to another state, with a different energy and momentum (this is schematically shown in Fig. 3.8). This mechanism is not very important for the heavy elements, for which photoionization is much more important. But free–free absorption is important for hydrogen and helium. Remember that since these are fully ionized, they do not contribute to photoionization; the atoms have been completely stripped of the electrons.

H. A. Kramers was the first to calculate the coefficient of absorption for such a free–free scattering.

$$a_{ff} = (....)\frac{1}{\nu^3}\frac{1}{\text{velocity}} \tag{3.19}$$

He showed that the absorption coefficient is, once again, inversely proportional to the cube of the frequency. Notice that in Eq. (3.19) the velocity of the electron comes in the denominator. This is as it should be. The faster the electron, the less it would be scattered by the nucleus. The above absorption coefficient varies smoothly with frequency, and *there is no restriction on the frequency as there was for photoionization.*

D. Scattering by electrons

You will notice that in the above absorption process, we were careful to state that an electron absorbed (or emitted) a photon while in the *vicinity* of an atomic nucleus (in other words, as it sweeps past it). You may have wondered why a lone electron cannot absorb energy from a photon. This has to do with Einstein's *Special Theory of Relativity*.

Let us try to understand this. Assume for a moment that a free electron absorbs energy from a photon. As a consequence, its kinetic energy and therefore its speed must increase. But different observers (in different states of motion) would not necessarily agree that this has happened. Indeed, to some observers it might appear that the speed of the electron has decreased! Surely, if the electron has really absorbed a photon, then *all* observers, regardless of their *frame of reference,* would *have* to agree that the kinetic energy of the electron has increased. Another way of stating this inconsistency is to say that a solitary electron cannot absorb a photon. If you like, the atomic nucleus in whose proximity the absorption of radiation takes place provides a frame of reference with respect to which all observers will agree that the kinetic energy of the electron has increased.

So what *can* a solitary electron do? What it can do is to *scatter the radiation.* In this process, an incident photon of frequency ν_1 is replaced by an outgoing photon of frequency ν_2, and the kinetic energy of the electron changes by $h(\nu_1 - \nu_2)$. *The kinetic energy can increase or decrease depending upon whether* $\nu_1 > \nu_2$ *or otherwise.* This is known as *Compton Scattering.* There is no problem with this as long as the photon is scattered in a direction different from the incident direction. In most situations this change in direction is more pronounced than the change in the frequency, and hence this process is called *scattering.* It must be borne in mind, however, that if the electron itself is moving with a speed close to that of light then the *gain* in the energy of photon can be spectacular. There are many astrophysical situations where radio waves are scattered by relativistic electrons. In these situations, the scattered photon can gain an enormous amount of energy from the electron. For example, radio waves can be *boosted in energy to* such an extent that they will manifest as *X-rays or even gamma rays!*

Let us return to the case of a stationary electron scattering a photon. It is easy to visualize this scattering process in classical physics. Let electromagnetic radiation be incident on a stationary electron (see Fig. 3.9). Recall that an electromagnetic wave is a transverse wave, characterized by oscillating electric and magnetic

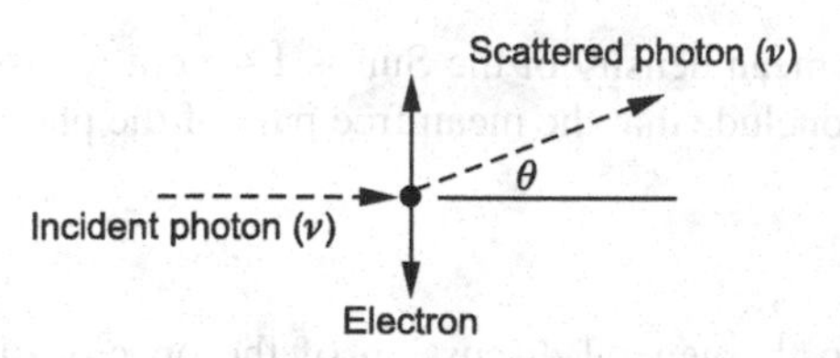

Fig. 3.9 The scattering of a photon by a stationary electron can be described using classical physics. The oscillating electric field associated with the incident electromagnetic wave sets the electron into oscillation. The electron will oscillate with a frequency which is equal to the frequency of the wave. According to electrodynamics, an oscillating electron will radiate at a frequency equal to the frequency of oscillation, but not necessarily in the direction of propagation of the incident wave. This is why this process is referred to as *scattering*

fields in a plane perpendicular to the direction of propagation. Let us concentrate on the electric field associated with the radiation. The electron will begin to oscillate under the influence of the oscillating transverse electric field of the radiation. Maxwell's theory of electrodynamics tells us that such an oscillating electron will radiate. *The frequency of the radiation emitted by the electron will be same as the frequency of its oscillation which, in turn, will be same as the frequency of the incident radiation.* This scattering process is known as *Thompson scattering*. Conservation of energy tells us that the scattered radiation is produced at the expense of the incident radiation. Classical physics tells us that the *cross-section* for this process per electron is:

$$\sigma_{\mathrm{Th}} = \frac{8\pi}{3}\frac{e^4}{m^2c^4} \tag{3.20}$$

This is known as the *Thompson scattering cross-section*. It is interesting to note that this is *independent of frequency* and is solely determined by fundamental constants! Imagine that radiation incident on a screen. The fraction of radiation absorbed due to this process per square centimetre will be equal to σ_{Th} times the number of electrons in the screen per unit volume (refer to the discussion of the mean free path, etc. at the beginning of this chapter). Since in this section we are using the terminology of opacity, rather than cross-section, let us have a feel for the opacity due to Thompson scattering. Recall that κ and σ are related through the simple relation:

$$\kappa\rho = \sigma n, \tag{3.21}$$

where ρ is the density and n is the number of electrons per cubic centimetre.

Since the cross-section is given in terms of fundamental constants, we can estimate the opacity. The mass absorption coefficient or opacity due to electron scattering is roughly equal to *0.4 (per square centimetre per gram of matter)*. At high temperature and pressure, when matter is likely to be fully ionised, electron scattering is the most important mechanism contributing to the opacity. Recall that the mean free path

$l = 1/\kappa\rho$. Since the mean density of the Sun is 1.4 g cm^{-3}, using the above value for the opacity, we conclude that the mean free path of the photon is approximately 2 cm.

4. Mean opacity

Let us now return to the general discussion of the opacity of stellar matter. We have discussed four mechanisms: photoionization; bound–bound transitions; free–free transitions; and electron scattering. The net absorption coefficient will be the sum of these opacities:

$$\kappa(\nu) = \kappa_{bf}(\nu) + \kappa_{bb}(\nu) + \kappa_{ff}(\nu) + \kappa_T . \tag{3.22}$$

We are not done yet. The absorption coefficient defined above is a function of the frequency of radiation; only the contribution to the opacity due to Thompson scattering is independent of frequency. But in our discussion of radiative equilibrium we have assumed a single value for the absorption coefficient throughout the star. Clearly, such a constant must represent some sort of *mean* over all the frequencies. In the 1920s, an astronomer named Rosseland gave a prescription for how such a mean is to be calculated, and that is still used. One must also bear in mind that the relative importance of the four mechanisms depends on the density and temperature. Therefore, calculating stellar opacity as a function of frequency, density and temperature is a big industry. But let us not get into those details!

5. The rate of energy production

The last thing one needs to know to construct a detailed model of the stars is the rate at which energy is produced. No one had a clue about this in the 1920s. Astronomers were still arguing about the mechanism of energy production. Eddington, of course, was convinced that the source of energy was the transmutation of hydrogen into helium. As we shall discuss in Chap. 5, *Energy Generation in the Stars,* a detailed theory of such fusion reactions was worked out only in 1938. Nevertheless, Eddington was able to make spectacular predictions based on his theory.

Let us now summarize what we have reviewed so far in this chapter. Having introduced the concept of *radiative equilibrium,* we set up the four equations of stellar structure (Eqs. 3.7–3.12). To solve for various properties of the stars one has to solve these equations self-consistently. The main inputs needed to solve these are:

1. An equation of state
2. The chemical composition
3. Opacity of the stellar material
4. The rate of energy production

For the equation of state, Eddington assumed that the gas can be described by Boyle's law. As for the chemical composition, he took the point of view that it really does not matter(!) as long as the composition was predominantly the heavy elements. He was so convinced that the interior temperature would be of the order of ten million degrees, and consequently most of the elements would be fully ionized, that he assumed that the dominant source of opacity would be electron scattering. The absorption coefficient due to this process does not involve any detailed calculation; it is determined by a combination of fundamental constants.

As we shall presently see, Eddington was able to make some remarkable predictions although he did not know the details of the energy production. It was as though Nature was in a conspiracy with him!

Eddington's Mass–Luminosity Relation

One of the remarkable predictions of Eddington's theory is regarding the luminosities of stars. Let us return to Eq. (3.12). You will recall that the condition of radiative equilibrium demands that the luminosity crossing an imaginary surface inside the star is given by:

$$L(r) = \text{(net outward flux through this surface)} \times \text{area}$$

$$L(r) = -\frac{ac}{3\kappa\rho}\frac{dT^4}{dr}4\pi r^2. \tag{3.23}$$

This is an astonishing result. It says that if we know the temperature gradient and the opacity we can predict the luminosity of a star.

Lets us try to estimate the luminosity of the Sun using this formula. Consider a point somewhere midway between the centre and the surface. In this case, r is equal to half the radius of the star. For the temperature that obtains there, let us assume ten million degrees (as per our earlier estimate). Let us replace the temperature gradient by

$$\frac{dT}{dr} \approx \frac{(T(r) - T_{\text{surface}})}{R},$$

where T_{surface} is the temperature at the surface and R is the radius of the Sun. For the opacity κ we shall assume that electron scattering is the dominant mechanism. As mentioned earlier, the opacity arising from Thompson scattering is approximately $0.4\,\text{cm}^2$ per gram of matter. If we now substitute for the average density of the Sun ($1.4\,\text{g/cm}^3$) we obtain for the reciprocal of $\kappa\rho$ approximately 2 cm (this is just the mean free path for the photons). Making these series of oversimplified assumptions we obtain from Eq. (3.12):

$$L \approx 3 \times 10^{35}\ \text{erg s}^{-1}.$$

You may be disappointed because this is about a hundred times larger than the luminosity of the Sun. But wait! Just think of the series of simplifying assumptions we have made. Besides, we know nothing about the chemical composition of the star, its internal temperature, the internal source of energy, the temperature gradient etc. And yet, we have obtained a value for the luminosity which is well within the range of stellar luminosities! *We have constructed a hypothetical body in which gravity is balanced by the combined pressure of the gas and radiation. The luminosity of such a hypothetical object could be anything. But extraordinarily, it turns out to be of the order of stellar luminosity*! This is the magic I was referring to earlier. We shall examine this more closely in the next chapter. But for now, let us proceed.

In making the above estimate, we made no reference to the mass of the star. Since in the ultimate analysis we expect gravity to be the source of all forms of energy, the mass of the star must surely be one of the factors that determine the luminosity generated. Let us now see what prediction Eddington's theory makes regarding this. Let us start with the condition of hydrostatic equilibrium given by Eq. (3.7):

$$\frac{dP}{dr} = -\frac{GM(r)\rho}{r^2}.$$

Like we did while estimating the luminosity, we shall replace all *differentials* by *differences*. Thus,

$$\frac{dP}{dr} \propto \frac{P}{R}. \tag{3.24}$$

For the characteristic density, we shall take

$$\rho \propto \frac{M}{R^3}. \tag{3.25}$$

Using relations (3.24) and (3.25), the equation of hydrostatic equilibrium gives the following scaling relation:

$$\boxed{P \propto \frac{M^2}{R^4}.} \tag{3.26}$$

You will recall that in Eddington's theory the pressure (the left-hand side of the above relation) that supports gravity is the sum of gas pressure and radiation pressure:

$$P = p_g + p_R$$

We shall now make the assumption that gas pressure dominates over radiation pressure and that the gas can be assumed to be a perfect (or ideal) gas, i.e.

$$P \approx p_g = \frac{\rho k_B T}{\mu m_H}. \tag{3.27}$$

It will turn out that this is a very good approximation for stars in the lower mass range (like the Sun, for example). Given this assumption we may write:

$$P \approx p_g \sim \frac{MT}{R^3}.$$

Introducing this into Eq. (3.26) we obtain the following interesting relation:

$$\boxed{RT \propto M.} \tag{3.28}$$

We are now all set to find the relation between the luminosity and the mass. Let us go back to our basic condition (3.23) for a star to be in radiative equilibrium, namely:

$$\boxed{L(r) = -\frac{ac}{3\kappa\rho}\frac{dT^4}{dr}4\pi r^2}.$$

Let us consider our imaginary surface to be the surface of the star. Therefore, $r = R$, the radius of the star. Making the usual approximation for the derivative on the right-hand side ($dT^4/dr \sim T^4/R$) we obtain:

$$L \propto \frac{(RT)^4}{M}.$$

Since $RT \propto M$, as may be seen in Eq. (3.28), we find the desired result.

$$\boxed{L \propto M^3} \tag{3.29}$$

This is the famous mass–luminosity relation derived by Eddington in 1924. And it is a remarkable result. It says that the luminosity of a star will be proportional to the cube of its mass. Notice that the radius of the star has dropped out of the equation. One would think that given a star of a certain mass, the luminosity it generates should depend on its radius. After all, the internal temperature should be determined by the radius—common sense tells us that smaller the star, the hotter it would be—and that, in turn, should determine the rate of energy generation.

It is almost as though the star knows the radius it must attain. Well, the principle of radiative equilibrium *dictates* to the star the luminosity it is allowed to generate. And that luminosity is determined only by its mass and the opacity [notice that the opacity enters the expression for the luminosity in Eq. (3.23)]. Given the opacity of stellar material, the star will adjust itself to that combination of RT such that *the energy generated per unit time precisely compensates for the heat energy lost from the surface per unit time.* If it were to generate more luminosity, given that the rate at which the energy can diffuse outwards is determined by the opacity, there will be a build up of energy in the interior and the condition of radiative equilibrium will be violated.

Comparison with Observations

Having obtained this relation between the mass and luminosity in 1924, Eddington went on to compare this prediction with observations. This is not as straightforward as it sounds. There are two practical difficulties. First, we should be able to estimate

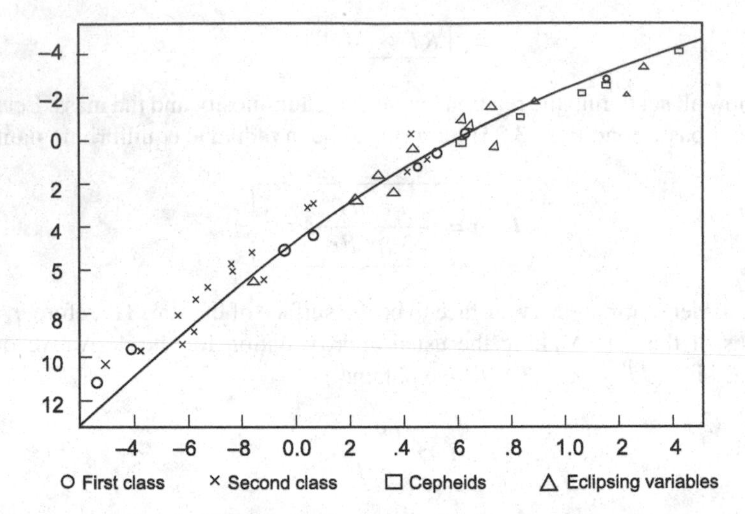

Fig. 3.10 The mass–luminosity relation reproduced from Eddington's *Internal Constitution of the Stars,* published by Cambridge University Press (1926). The logarithm of the masses of the stars is plotted on the x-axis. Plotted on the y-axis is the *absolute bolometric magnitude* of the stars. This is a measure of the luminosity of the stars. The luminosity increases vertically along the y-axis; the smaller the absolute magnitude, larger is the luminosity. The *curve* is the theoretical mass–luminosity relation derived by Eddington: equation (3.29)

the *luminosity* or *intrinsic brightness* of the star from its apparent brightness. This is like trying to determine whether a distant street light has a 100 W bulb or a 500 W bulb. We can deduce this from the apparent brightness provided we know the distance to the lamp. And so it is with the stars. Unfortunately, until very recently, one could accurately estimate the distance of only a handful of stars.

Next, we have to know the *mass* of the star. This is also a tricky business. There are special classes of stars whose mass can be determined with reasonable accuracy. Binary stars belong to this class. Such systems consist of two stars going around a common centre of mass. As the stars go around in their orbits, the observed wavelength of known spectral lines from the stars will undergo periodic changes due to the well known Doppler Effect. The observed wavelength will be shorter when the star is approaching us (blue shift), and will be longer when the star is receding from us (red shift) during its revolution in its orbit. By studying this periodic modulation in the received wavelength of spectral lines one can determine the orbital period and the semi-major axis of the orbits. One can then use Kepler's laws to estimate the masses.

Figure 3.10 shows the comparison Eddington made in 1924 between his theory and the observationally deduced luminosities of stars of known mass (reproduced from Eddington's classic book *The Internal Constitution of the Stars*). A few words of explanation are in order. The y-axis is the luminosity of the stars in units which astronomers prefer to use (*Absolute Bolometric Magnitude*).We shall not go

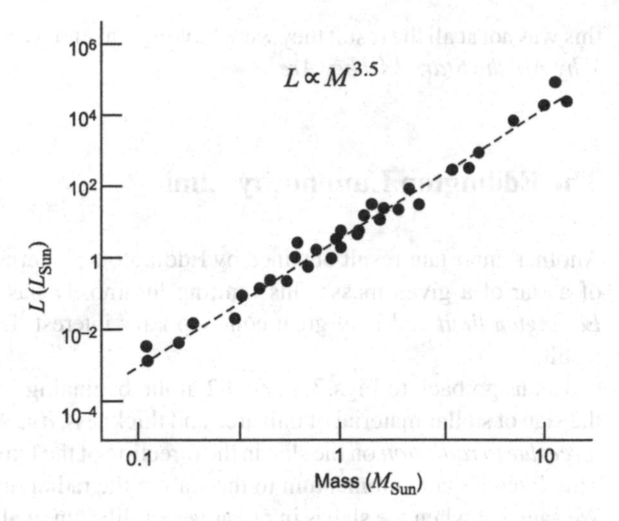

Fig. 3.11 A log–log plot of the mass–luminosity relation using more recent data. As will be seen, an exponent of 3.5 fits the data very well. [mass–luminosity relation. *The Columbia Encyclopedia, Sixth Edition 2008*. Retrieved July 20, 2009 from Encyclopedia.com: http://www.encyclopedia.com/doc/1E1-masslumi.html]

into it here, except to say that the luminosity increases vertically; as the numbers decrease the luminosity increases. Conversely, as the numbers increase the luminosity decreases.

As Eddington put it, the *magnitude scale is like golfer's handicap—the bigger the number the worse the performance!* The brightest star, at the top right-hand corner of Fig. 3.10, is more than two million times more luminous than the faintest star, at the left bottom. The x-axis is the logarithm of the mass of the star in units of the mass of the Sun, i.e., $M/M_{\odot}$. At the extreme left is a star with a mass approximately one-sixth the mass of the sun, and the mass of the star on the extreme right is about 30 $M_{\odot}$; extraordinarily, there are very few stars outside this range of masses. The curve is the theoretical mass–luminosity relation given by Eq. (3.29), and the observational data are the various symbols marked.

It is important to note that in any theoretical formula there will be a number of parameters that one cannot accurately specify. In the present case, for example, even though one may know the mass of a star it is difficult to accurately predict its luminosity because one or more numerical constants appearing in the formula cannot be determined with any confidence from pure theory. In such circumstances the usual practice is to make the curve *fit* one data point; this would then serve to determine the *unknowns* in the formula. While attempting to compare his theory with observations, Eddington made his curve pass through one data point. He chose the bright star *Capella*. The theoretical curve has thus been *normalized*, and it cannot be raised or lowered to force it to *go through* other data points. You will agree that the agreement between observations and the theory is nothing less than spectacular.

More recent observational data are shown in Fig. 3.11. As will be seen, $L \propto M^{3.5}$ gives an excellent fit to this data set. This slope is very nearly what Eddington's theory predicts. The agreement of the observations with the theoretical curve came as a complete surprise to everyone, including Eddington. It was a surprise because this was not at all the result they were looking for, and we shall discuss this in Chap. 4, *Why Are the Stars As They Are.*

The Eddington Luminosity Limit

Another important result obtained by Eddington concerns the maximum luminosity of a star of a given mass. This limiting luminosity has come to be known as the *Eddington limit* and is of great contemporary interest. Let us derive this important result.

Let us go back to Figs. 3.1 and 3.2 at the beginning of this chapter and consider the slab of stellar material of unit area and thickness, dr. We argued that the *resultant force due to radiation* on the slab in the direction of the temperature gradient is $-dp_R$. This force imparts momentum to the slab in the radial direction equal to $F\kappa\rho dr/c$. We said that when the slab is in radiative equilibrium, it absorbs this momentum flux and uses it to supplement the gas pressure in supporting itself against the inward directed force of gravity.

Let us now assume that gas pressure is negligible compared to radiation pressure. Under the circumstances when a star is generating the maximum possible luminosity, it is quite reasonable to assume that radiation pressure dominates over gas pressure. In this case, the condition of equilibrium requires that the outward-directed resultant force due to radiation precisely balances the inward-directed force on the slab due to gravity. In other words,

$$-dp_R = \frac{GM(r)\rho(r)dr}{r^2}. \tag{3.30}$$

In Eq. (3.30), $M(r)$ is the mass interior to the slab and $\rho(r)dr$ is the mass of the slab (remember it has unit cross sectional area and thickness dr). Using the relation

$$-dp_R = \frac{F\kappa\rho dr}{c},$$

the condition for equilibrium can be recast as

$$\frac{F\kappa\rho dr}{c} = \frac{GM\rho dr}{r^2}. \tag{3.31}$$

Let us rewrite this in terms of the *luminosity* L, which is simply the net outward flux F crossing a unit area multiplied by the area of the sphere ($L = 4\pi r^2 F$). Therefore,

$$\frac{L\kappa\rho dr}{4\pi r^2 c} = \frac{GM\rho dr}{r^2}. \tag{3.32}$$

Simplifying we get

$$\boxed{L = \frac{4\pi c G M}{\kappa}.} \tag{3.33}$$

Remember that the expression we have just derived represents the limiting luminosity the star can generate. If the luminosity exceeded this then the outward directed force due to the radiation would be greater than the force due to gravity, and the star will be blown apart. Notice that this limiting luminosity is uniquely determined by the mass of the star and the opacity (or the mass absorption coefficient). Let us now assume that hydrogen is the predominant element and that the stellar matter is completely ionized. Under these conditions, *electron scattering* will be the ultimate source of opacity. So κ in the above formula would be the *Thompson absorption coefficient.* κ_{Th}. It is customary to recast the formula for the limiting luminosity in terms of the *Thompson scattering cross-section* σ_{Th}. Remember that κ and σ are related by the relation $\kappa\rho = \sigma n$, where ρ is the mass density and n is the number of free electron per unit volume. Since we are dealing with hydrogen, $\rho = n m_H$, where m_H is the mass of the hydrogen atom.

Let us now rewrite the expression for the limiting luminosity as

$$\boxed{L_{\mathrm{Eddington}} = \frac{4\pi c G M m_H}{\sigma_{\mathrm{Th}}}.} \tag{3.34}$$

This limiting value of the luminosity has come to be known as the *Eddington Luminosity Limit.* It is a truly remarkable result. No matter how efficient the process of energy generation might be, a self gravitating body of mass M cannot generate luminosity greater than this, for if it did, radiation pressure will overwhelm gravity and the object would become unstable. It is also a beautiful result. Apart from fundamental constants, *the critical luminosity is uniquely determined by the mass of the body* (remember that the Thompson cross-section only involves e, m_e and c).

It is useful to have an idea of the numerical value of the *Eddington limit.* Substituting for the constants and the mass of the Sun, we obtain

$$L_{\mathrm{Edd}}(\mathrm{Sun}) = 3.3 \times 10^4 L_\odot \approx 10^{38}\,\mathrm{erg\,s^{-1}}. \tag{3.35}$$

Since this result is of great significance is many situations, it is useful to write it as follows

$$\boxed{L_{\mathrm{Edd}} = \left(\frac{M}{M_\odot}\right) \times 10^{38}\,\mathrm{erg\,s^{-1}}.} \tag{3.36}$$

As we shall see in the later volumes in this series, this result is of significance in the context of X-ray astronomy, accretion discs around black holes, etc. Just to get a feeling for the importance of this concept, I shall mention one example without attempting to really explain it in detail. When the so called *quasars* were discovered in the 1960s, astronomers were astonished. Although there was direct observational evidence to suggest that these objects—whatever they might be—were very compact (just a few *light days* across), their luminosities were of the order of 10^{46} erg s^{-1} (truly staggering). The Eddington limit was invoked to argue that *an object radiating* approximately 10^{46} erg s^{-1} *must be more massive than* $10^8\ M_\odot$; *otherwise, radiation pressure would disrupt it.* Such a large mass, packed into a small volume, immediately suggested that the central engines of quasars must be *supermassive black holes!*

On that note, we shall wind up this chapter. This chapter has been devoted to Eddington's pioneering efforts to construct a comprehensive theory of the stars. We have only touched upon the essence of his theory. By choosing to highlight the mass–luminosity relation and the critical luminosity, we have attempted to bring out the *magical* elements of his theory. But there is much more than that in his classic book *The Internal Constitution of the Stars* published by Cambridge University Press in 1926. It is important to appreciate that this book was not a survey of the published literature. Much of it is pioneering and original contribution by Eddington. The number of insights and prescient conjectures that are to be found in this book are truly impressive:

1. The idea that radiation pressure must play an increasingly important role in the stability of the stars as one goes to more massive stars.
2. The concept of radiative equilibrium, which we have repeatedly stressed.
3. The idea that the source of energy in the stars must be the transmutation of hydrogen into helium.
4. The realisation that photoionization must be a very important source of opacity.
5. The mass–luminosity relation.
6. The concept of a limiting luminosity.

To really appreciate his monumental efforts one has to study—and admire—his book. To me, two things stand out. First, the remarkably lucid style in which it is written. His examples and analogies put one at ease. This should be apparent from the number of extended quotations I have included from Eddington's book. Second, the rapidity with which he absorbed the latest developments in physics and used them in an astronomical context is simply unparalleled.

Soon after it was published, the famous Princeton astrophysicist Henry Norris Russell wrote, '(the book) has every claim to be regarded as a masterpiece of the first rank'. Few books have had as much influence in the development of a subject as this book has had. More than 80 years after it was published, it continues to be regarded a *masterpiece of the first rank.*

Chapter 4
Why are the Stars as they are?

Are Stars Really Globes of Perfect Gas?

As we saw in the previous chapter, Eddington's theory of stars, based on the principle of radiative equilibrium, was spectacularly successful in predicting their *mass–luminosity relation*.

Recall that Eddington's theory was predicated on two assumptions. First, the star is in radiative equilibrium. Second, the material constituting the interior of a star can be described as a *perfect gas*. When Lane proposed his theory of the stars on the assumption that they were blobs of perfect gas, there was no evidence that any such star existed. In 1870, *the only star whose average density one knew was the Sun* ($1.4\,g/cm^3$). In our experience on the Earth, gases cease to behave as a perfect gas long before they reach a density $\sim 1\,g/cm^3$. Water vapour, for example, becomes *liquid water* when compressed to a density of $1\,g/cm^3$. Therefore, Lane's assumption seemed quite unreasonable. By 1924, it was well established that some stars at least were quite diffuse. By then, the diameters of some giant stars had been measured, allowing one to deduce their mean density. For example, the mean density of *Capella* (at the extreme right of the mass–luminosity diagram shown in Fig. 3.10) is less than the density of air. In fact, this is true for all the stars in the right-hand-side half of that figure. For these diffuse stars, perfect-gas behaviour can safely be assumed. But the stars on the left-hand side of the diagram are much denser than water. The star *Krueger 60*, for example, is denser than iron!

You will therefore appreciate that there was total surprise when the data on the luminosities of real stars agreed so well with the predictions of a theory which assumed that stars are blobs of *perfect gas*. To Eddington, this *agreement* was most annoying! He exclaimed, 'What business have they on a curve reserved for a perfect gas?' Eddington had hoped for something different. The luminosities of the *dense stars* were expected to be well below the theoretical curve for diffuse stars. Astronomers had hoped to learn something about the stars by observing 'how far the dense stars fell below the theoretical curve'. This is how science progresses. The 'deviations' from the predictions of a successful theory are the stepping stones for further progress.

G. Srinivasan, *What are the Stars?* Undergraduate Lecture Notes in Physics,
DOI: 10.1007/978-3-642-45302-1_4,

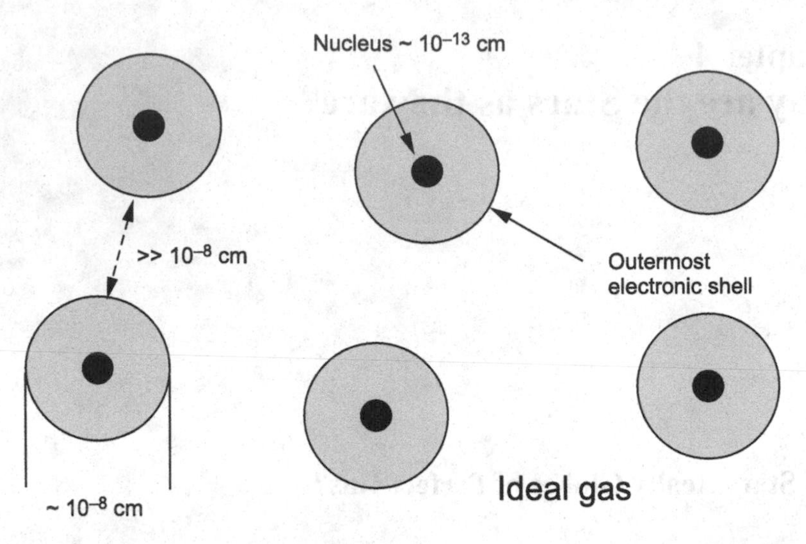

Fig. 4.1 A gas is considered to be *ideal* or *perfect* if the kinetic energy of the atoms is much greater than the potential energy of interaction between two atoms. In a *classical gas* made of atoms or molecules, this condition will be satisfied if the mean distance between the atoms is much larger than the characteristic size of the atoms

What should we make of this agreement between the prediction for stars made of *perfect gas* and the data pertaining to real stars? Is it possible that the stellar material of even the dense stars is perfect gas? *But this would require that gas of density far in excess of the density of iron would behave as a perfect gas.* As we shall now argue, this is certainly possible. To quote Eddington, 'the reason why it should not *is earthly* and does not extend to the stars!'

Let us try to understand this remarkable statement. The feature of a perfect gas obeying Boyle's law is that the distance between the atoms is very large compared with the size of the atom (see Fig. 4.1).

This is why the potential energy of interaction between the atoms is small compared with their kinetic energy; recall that this is the condition for a gas to be regarded as *perfect* or *ideal*. As Eddington would put it, a gas contains very little substance and a lot of emptiness. As the gas is compressed, some of the empty space is squeezed out and the atoms come closer. At some stage the force between the atoms becomes significant to affect the *compressibility* of the gas. When this happens, the gas ceases to be a perfect gas; Boyle's law breaks down (see Fig. 4.2).

As we compress the gas further the atoms will be practically touching one another. When the density reaches a few grams per cubic centimetre the gas will become a liquid. Let us recall that the characteristic size of a neutral atom is approximately 10^{-8} cm. Strictly speaking, this is the size of the hydrogen atom, with only one electron. In heavier atoms, the cloud of electrons will be a little bigger than this; but 10^{-8} cm is the typical size of most atoms. Since the atoms are very close to their neighbours, the interaction between the atoms is very strong in a liquid. If we

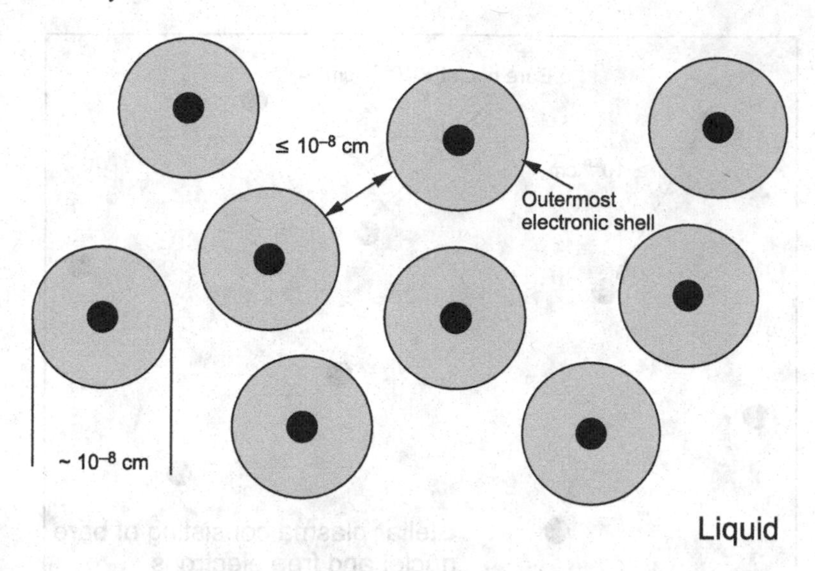

Fig. 4.2 A classical gas ceases to be *ideal* when the distance between the atoms/molecules becomes comparable to their sizes, and consequently the interaction energy between the particles becomes important. At a density of a few grams per centimetre cube, most gases will undergo a phase transformation to a liquid. In a liquid the atoms will be practically touching one another, but they would still be in a state of motion like in a gas. At higher densities, these motions will cease, and the liquid solidifies. The atoms in a solid can vibrate about their mean positions, but they cannot wander around

squeeze the atoms even closer to one another, the liquid becomes a solid. The reason for this is simple. The repulsion between the atoms will be very strong when they are practically touching each other. The best way for the atoms to *minimize* this interaction energy is to *stay put in one place*, and that is what a solid is.

The point to note is that neutral atoms of size $\geq 10^{-8}$ cm do not exist in the stars (see Fig. 4.3). As we have already discussed, at the conditions of enormous temperature and pressure that prevail in the interior of stars, stellar atoms will be stripped of their electrons. The lighter atoms will be stripped right down to the nucleus. Recall that the size of the nucleus (approximately 10^{-13} cm) is approximately a hundred thousand times smaller than the size of the neutral hydrogen atom. The heavier atoms will retain a few electrons in the innermost orbits, but the ion will still be a thousand times smaller than the neutral atom with its full compliment of electrons. The upshot of all this is the following. *Even at densities much greater than that of iron, the mean distance between the bare nuclei of the stellar atoms would still be enormous compared to their size.* With their electronic shells stripped off, the stellar atoms are incredibly small compared to their terrestrial counterpart. It is therefore not a surprise that the stellar material remains a perfect gas even at very high densities. The key to this is, of course, the very high temperature and pressure that *obtain* in the stars. Clearly, such extreme conditions are not *earthly*.

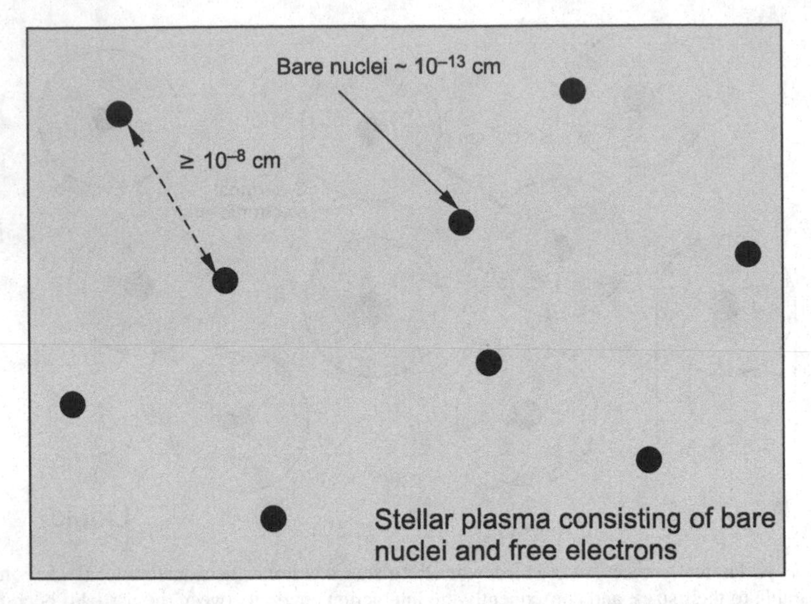

Fig. 4.3 Neutral atoms do not exist at the conditions of high temperature and pressure that prevail in the stars. The light elements will be completely ionized and the atoms of the heavy elements will retain only the electrons in the innermost shells. Since the atomic nuclei are roughly a hundred thousand times smaller than the neutral atoms, even at a density of a few grams per centimetre cube the distance between the bare nuclei would be enormous compared to their sizes. Therefore, the stellar plasma can be regarded as *ideal* even at the density of iron!

Fig. 4.4 When *crinolines* were fashionable, a ballroom with just a handful of dancers would be as congested as a jam-packed modern day disco! Even though the number of dancers would be very few, their effective sizes would be large

We should have expected this. Eddington put it in his characteristic fashion: '*Our mistake was that in estimating the congestion in the stellar ball-room we had forgotten that crinolines are no longer in fashion*'. Since ballroom dancing, with women wearing *crinolines,* belongs to a bygone era, let us consider a more familiar situation (Fig. 4.4). Imagine a very crowded street on a rainy day with every pedestrian carrying a gigantic open umbrella. The umbrellas will be rubbing against each other. The people with these umbrellas are our terrestrial atoms; the pedestrians are the nuclei and the umbrellas the electron shells. Imagine the rain stops, and everyone folds their umbrellas. Suddenly there will be a lot of space between the people, and they can move with relative ease. But the density of people has not changed!

Returning to the mass-luminosity relation derived by Eddington, we need not be concerned. We have *not* compared the theory with the *wrong stars* after all. *The stellar material in all the stars, diffuse as well as dense, can be safely regarded as perfect gas.*

The Happening of the Stars

We turn next to the *happening of the stars*. We can do no better than quote Eddington himself. 'It is remarkable that the units into which the matter of the universe has aggregated primarily are so nearly alike in mass. The stars differ widely in brightness, density and physical conditions, but they mostly contain the same amount of material. It is as though nature had a standard model before her in forming the stars, and (except for occasional lapses of vigilance) would not tolerate much deviation. The extreme range (about one-sixth to a hundred times the Sun's mass) does not give a fair idea of the general uniformity of mass. ... A mass range of 5:1 would, I believe, include more than 90 % of the stars'. (Source: *The Internal Constitution of the Stars*).

Eddington went on to *explain* the reason for this in his famous parable of a physicist in a cloud-bound planet. It is such an extraordinary explanation that this book would not be complete without narrating it. We quote Eddington:

> The outward-flowing radiation may thus be compared to a wind blowing through the star and helping to distend it against gravity. The formulae to be developed later enable us to calculate what proportion of the weight of this material is borne by this wind, the remainder being supported by the gas pressure. To a first approximation, the proportion is the same in all parts of the star. It does not depend on the density or on the opacity of the star. It depends only on the mass and molecular weight. Moreover, the physical constants employed in the calculation have all been measured in the laboratory, and no astronomical data are required. We can imagine a physicist on a cloud-bound planet who has never heard tell of the stars calculating the ratio of radiation pressure to gas pressure for a series of globes of gas of various sizes, starting, say, a globe of mass 10 gm, then 100 gm, 1000 gm, and so on, so that the *n*th globe contains 10^n gm. Table 4.1 shows the more interesting part of his results.
>
> The rest of the table would consist mainly of long strings of nines and zeros. Just for the particular range of mass about the thirty-third and the thirty-fifth globes the table becomes interesting, and then lapses back to the nines and zeros again. Regarded as a tussle between

Table 4.1 Radiation pressure and gas pressure corresponding to globe size

No. of globe	Radiation pressure	Gas pressure
30	0.00000016	0.99999984
31	0.00016	0.999984
32	0.0016	0.9984
33	0.106	0.894
34	0.570	0.430
35	0.850	0.150
36	0.951	0.049
37	0.984	0.016
38	0.9951	0.0049
39	0.9984	0.0016
40	0.99951	0.00049

matter and ether (gas pressure and radiation pressure) the contest is overwhelmingly one-sided except between No. 33–35, where we may expect something interesting to happen.

What *happens* is the stars.

We draw aside the veil of cloud beneath which our physicist has been working and let him look up at the sky. There he will find a thousand million globes of gas nearly all of mass between his thirty-third and thirty-fifth globes: that is to say, between one-half the Sun's mass and fifty times. The lightest known star is about 3×10^{32} g. and the heaviest about 2×10^{35} g. The majority are between 10^{33} and 10^{34} g. where the serious challenge of radiation pressure to compete with gas pressure is beginning.

What Eddington meant was this. The masses of the stars are as they are because only then radiation pressure will be comparable to gas pressure (see the Table 4.1). This is so only in a very narrow range of mass. And this tussle between gas pressure and radiation pressure is necessary for their *happening*. Why this should be so is not at all clear. We shall now try to understand this mysterious statement.

Why are the Stars as they are?

In the earlier chapters, we discussed the question, *What are the stars?* We turn next to an even more interesting question, namely, *Why are the stars as they are?* It is a remarkable fact that most of the stars have masses in an incredibly narrow range of mass. As we just saw, Eddington offered a plausible explanation for this, although his explanation is obscure. There are two distinct things that need to be explained:

1. Why do the stars have *nearly the same mass?*
2. Why is the *tussle between gas pressure and radiation pressure* important for the happening of the stars?

Eddington is silent on both these questions. A much more satisfactory explanation was provided by Subrahmanyan Chandrasekhar. Before presenting his explanation, let us digress a little. Let us remind ourselves of what we are attempting to do. What we are really trying to do is to understand why Eddington's theory of the stars is so enormously successful.

Natural phenomena are often circumscribed by well-defined *scales*. These are scales of *length, time, mass* etc. Theories concerning natural phenomena are successful only to the extent that these scales naturally emerge in the theories. Let us consider the example of the *theory of hydrodynamics* and its description of waves in a liquid. Hydrodynamics is not concerned with the constituent atoms and molecules of the liquid. Its starting point is in defining the *macroscopic properties* of the liquid, such as its average density, average pressure, etc. The motions of the atoms/molecules, their collisions and so on are *averaged out* in defining the macroscopic properties mentioned above. Therefore, the subject of hydrodynamics cannot deal with phenomena whose length scale is comparable to the distance between the atoms, or whose time scale is comparable to the mean time between the collisions of the atoms. It can only hope to describe waves whose wavelengths are much larger than the distance between the microscopic particles and whose time periods are much longer than the time between collisions of the particles. Hydrodynamics will predict some characteristic scales of length and time, which in turn are determined by properties such as the mean density, pressure, gravity, and so on. The theory will be most successful in describing those waves whose wavelengths and periods are close to these naturally emerging length and time scales within the theory.

Let us consider another example from the realm of atomic physics. In analogy to the question we are addressing concerning the stars, let us ask, *why are the atoms as they are?*

What do we mean by such a question?

Atoms of the various elements of the periodic table, although they contain vastly different number of particles, are roughly the same size: a few times 10^{-8} cm. The important thing to appreciate is that while their sizes may vary by a factor of fifty or even a hundred, we do not see occurring in nature atoms as big as a ball bearing, let alone a ping pong ball. Why is this so? Let us try to understand this in a simple way.

As you know, atoms cannot be described by classical mechanics; the theory that is so successful in describing the planetary system. In 1913, Neils Bohr formulated his celebrated theory of the atoms. At first sight, this theory may appear no different from classical mechanics: he equated the electrostatic force experienced by the electron (due to the positive charge of the central nucleus) to the centrifugal force experienced by the electron due to its motion around the nucleus. This is exactly what one does in the case of planetary motion around the sun; the attractive force between the sun and the planets is due, of course, to gravity.

In addition to *balancing the forces*, like we do in classical physics, Bohr introduced another rule; a rule that was not there in classical mechanics but which has to be obeyed in the world of the atoms. He said that *the angular momentum of the electron must be an integral multiple of Planck's constant h*. (Angular momentum is the momentum related to the electron's circular motion around the nucleus.) When one introduces this new and essential principle, a length scale naturally emerges in the theory; a length scale which is defined by a specific combination of fundamental constants that appear in the theory of the atoms (Planck's constant h, the mass of the electron m_e and the charge of the electron e). This combination of the fundamental constants, which has come to be known as the *Bohr radius,* has the dimension of a length, and is given by the expression:

$$\boxed{\frac{h^2}{4\pi^2 m_e e^2} \sim 0.5 \times 10^{-8}\ \text{cm}} \tag{4.1}$$

The point to appreciate is this. There is no *a priori* reason why Neils Bohr's rule should apply in nature. There is no reason why such objects as atoms should exist obeying the rules stated by Bohr. But if they do exist, then their sizes would be comparable to the Bohr radius. To put it differently:

> Atoms are as they are because they obey the rules of atomic physics. And atomic physics provides a length scale with which to measure them, namely the Bohr radius.

Let us now return to the question, *why are the stars as they are?* In analogy with the above discussion, we have to ask whether Eddington's theory naturally isolates a combination of fundamental constants occurring in it with the dimension of mass; if so, what is its numerical magnitude? Interestingly, Eddington did not do this. This is particularly surprising because in the later part of his career Eddington was very preoccupied with natural constants. The discussion below is due to Chandrasekhar (1937).

After his monumental discovery concerning the ultimate fate of the stars (which will be the theme of the second volume in this series), Chandrasekhar turned to a detailed investigation of the internal constitution of the stars. He extended and developed the basic methods introduced earlier by the masters of the subject such as Eddington, Milne, and others. Along the way, he proved a number of mathematical theorems (which was to become characteristic of his work for the next sixty five years) concerning the equilibrium or *mechanical stability* of the stars. One of those theorems holds the key to our question.

A Characteristic Mass for the Stars

We shall now outline how Chandrasekhar's theorem isolates a *mass scale* for the stars. Let $\rho(r)$ be the density profile of a star of radius R (central panel of Fig. 4.5). Construct two other stars of uniform density. On the left is a star with a constant density equal to the *average density*, $\bar{\rho}$, and on the right is another star with a constant density equal to the *central density*, ρ_c. The radii of these configurations will be R and R_c, respectively. According to one of the theorems of Chandrasekhar, for a star to be stable the total pressure, P_c, at the centre of the actual star must satisfy the following inequality:

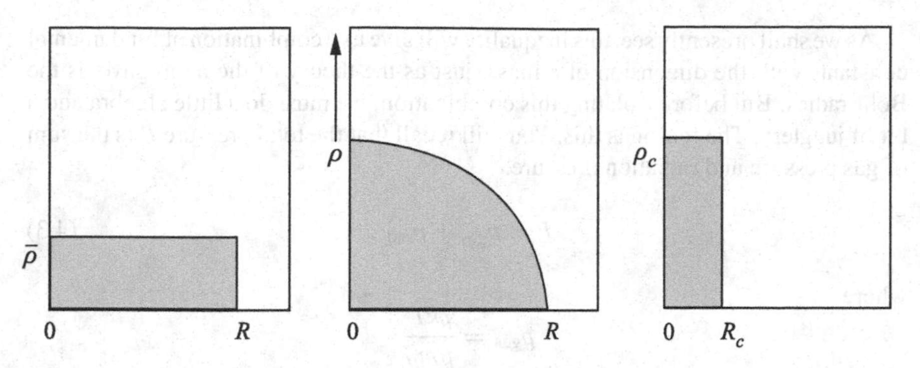

Fig. 4.5 The central panel shows the density distribution of the actual star. At the *left* and *right* are two *model stars*. On the *left* is a star of uniform density which is equal to the average density of our star. The radius of this model star will, of course, be equal to the radius of our star. On the *right* is another model star of uniform density equal to the central density of our star

$$\frac{1}{2}G\left(\frac{4\pi}{3}\right)^{1/3}\bar{\rho}^{\frac{4}{3}}M^{\frac{2}{3}} \leq P_c \leq \frac{1}{2}G\left(\frac{4\pi}{3}\right)^{1/3}\rho_c^{\frac{4}{3}}M^{\frac{2}{3}}. \tag{4.2}$$

Let us try to understand this theorem. At the centre of this inequality is the *total pressure* at the centre of a star, P_c, namely the sum of gas pressure and radiation pressure. On the left is the *gravitational pressure* at the centre of the model star with uniform density equal to the average density $\bar{\rho}$. On the right is the gravitational pressure at the centre of the star with uniform density equal to the central density, ρ_c:

Central Gravitational Pressure ($\bar{\rho}$) $\leq (p_g + p_R)_c \leq$ Central Gravitational Pressure (ρ_c).

You may be puzzled by the expression used above for the gravitational pressure because in Chap. 2, *Stars as Globes of Gas,* the expression GM^2/R^4 has been used for gravitational pressure. This can easily be recast as follows:

$$\frac{GM^2}{R^4} \Rightarrow \frac{GM^{2/3}M^{4/3}}{R^4} \Rightarrow GM^{2/3}\left(\frac{M}{R^3}\right)^{4/3} \Rightarrow GM^{2/3}\rho^{4/3}.$$

This is the way Chandrasekhar has expressed the gravitational pressure in his inequality theorem. You will see the reason for it presently.

Let us now return to the stability theorem. If this inequality is violated then there must be regions in the star where adverse density gradients prevail, and this will lead to instabilities. *Thus, satisfying this inequality is a necessary condition for the stable existence of the star.*

As we shall presently see, this inequality will give us a combination of fundamental constants with the dimension of a mass, just as the theory of the atom gave us the Bohr radius. But before isolating this combination, we must do a little algebra and a bit of jugglery. The reason is this. You will recall that the total pressure P is the sum of gas pressure and radiation pressure.

$$P = p_{\text{gas}} + p_{\text{rad}} \tag{4.3}$$

where

$$p_{\text{gas}} = \frac{\rho k T}{\mu m_p}$$

$$p_{\text{rad}} = \frac{1}{3} a T^4.$$

Note that while radiation pressure is determined by the temperature alone, gas pressure involves both the temperature and density. Therefore the total pressure is a function of *both* density and temperature. Since the left and the right sides of the inequality (4.2) involve only the density it is difficult to extract any meaningful conclusions from the inequality. We must therefore employ a little trick so that the temperature is eliminated from the expression for the total pressure. Only then we can meaningfully compare the two sides of the inequality.

Following Eddington, let us introduce a fraction, β, which is defined below:

$$P = \frac{1}{\beta}\left(\frac{\rho k T}{\mu m_p}\right) = \frac{1}{1-\beta}\left(\frac{1}{3} a T^4\right). \tag{4.4}$$

The meaning of β is clear. It is the *fraction of the total pressure contributed by gas pressure*, $(1-\beta)$ *is the fraction due to radiation pressure*. We can equate the second and the third terms in the equation above and express T in terms of β and ρ:

$$\frac{1}{\beta}\frac{\rho k T}{\mu m_p} = \frac{1}{1-\beta}\frac{1}{3} a T^4.$$

Solving for T we get:

$$\boxed{T = \left[\frac{3}{a}\frac{k}{\mu m_p}\frac{1-\beta}{\beta}\right]^{\frac{1}{3}} \rho^{\frac{1}{3}}.} \tag{4.5}$$

We can now use this expression to recast the total pressure in terms of ρ and β, instead of ρ and T. Let us consider the following equation expressing the gas pressure as a fraction of the total pressure:

$$P = \frac{1}{\beta}\frac{\rho k T}{\mu m_p}.$$

Let us substitute for T in terms of ρ and β using the expression given in Eq. (4.5). A little rearrangement gives:

$$\boxed{P = \left[\frac{3}{a}\left(\frac{k}{\mu m_p}\right)^4 \frac{1-\beta}{\beta^4}\right]^{\frac{1}{3}} \rho^{\frac{4}{3}}.} \tag{4.6}$$

This gives us an alternate expression for the total pressure in terms of ρ and β. We have not got rid of the temperature dependence; it is now implicit, rather than explicit.

Let us now return to Chandrasekhar's stability theorem (4.2) and use this expression for the total pressure expressed in terms of ρ and β:

$$\frac{1}{2}G\left(\frac{4\pi}{3}\right)^{\frac{1}{3}} \bar{\rho}^{\frac{4}{3}} M^{\frac{2}{3}} \leq \left[\frac{3}{a}\left(\frac{k}{\mu m_p}\right)^4 \frac{1-\beta_c}{\beta_c^4}\right]^{\frac{1}{3}} \rho_c^{\frac{4}{3}} \leq \frac{1}{2}G\left(\frac{4\pi}{3}\right)^{\frac{1}{3}} \rho_c^{\frac{4}{3}} M^{\frac{2}{3}}. \tag{4.7}$$

Now both the total pressure (the central term in the above inequality) and the gravitational pressure (the left and right sides of the inequality) have been expressed in terms of the density alone; the explicit dependence of the total pressure on the temperature has been replaced by an implicit dependence, through the use of the dimensionless fraction β. Notice that since it is the central value of the total pressure that enters the inequality, we have added a subscript to the dimensionless fraction, β. Since the density appears as $\rho^{4/3}$ on both sides of the inequality, it can be cancelled. That is the trick of introducing the fraction, β.

The right-hand part of the inequality now reads as follows:

$$\left[\left(\frac{k}{\mu m_p}\right)^4 \frac{3}{a}\frac{1-\beta_c}{\beta_c^4}\right]^{\frac{1}{3}} \leq \left(\frac{\pi}{6}\right)^{\frac{1}{3}} G M^{\frac{2}{3}}. \tag{4.8}$$

Remember that Stefan's constant, a, and Boltzman's constant, k_B, that enter the above equation are not fundamental constants. Stefan's constant is defined as:

$$a = \frac{8\pi^5 k^4}{15c^3h^3}.$$

Substituting this in the above equation and simplifying one gets:

$$\mu^2 M\left(\frac{\beta_c^4}{1-\beta_c}\right)^{\frac{1}{2}} \geq 0.19\left[\left(\frac{hc}{G}\right)^{\frac{3}{2}}\frac{1}{m_p^2}\right]. \tag{4.9}$$

In the above inequality, μ is the molecular weight of the stellar material ($\mu \sim 1$ if it is predominantly hydrogen and $\mu \sim 2$ if there is very little hydrogen. In the latter case it really does not matter which of the elements heavier than hydrogen is dominant) and β_c is the value of the dimensionless fraction at the centre of the star. We have now achieved our goal!

We observe that the above inequality (4.9) has isolated the following combination of fundamental constants that has the dimension of mass:

$$\boxed{[M] = \left(\frac{hc}{G}\right)^{\frac{3}{2}} \frac{1}{m_p^2}.} \tag{4.10}$$

Insert the dimensions of all the constants on the right-hand side and convince yourself that the combination does, indeed, have the dimension of mass. This is a famous combination of fundamental constants, and will play a very prominent role in the subsequent volumes. But for now, let us savour this remarkable result.

First, what is the significance of this result? Recall that in Bohr's theory of the atom, a combination of fundamental constants having dimensions of length naturally emerged, namely, the Bohr radius. This length defined the characteristic size of the atoms. Now we know that the actual sizes of the atoms are indeed comparable to the Bohr radius, thus confirming that Bohr's ideas are at the base of quantum physics. In a similar fashion, a characteristic mass emerges in Eddington's theory of the stars, a theory in which gravity is balanced by the combined pressure of the gas and radiation. What is the numerical value of this characteristic mass?

Inserting the values of the fundamental constants, we find that Eddington's theory of the stars gives the following *characteristic scale for the mass:*

$$\boxed{\left(\frac{hc}{G}\right)^{\frac{3}{2}} \frac{1}{m_p^2} \cong 29.2 M_{\odot}.} \tag{4.11}$$

Note that the numerical value of this mass scale is 29.2 times the mass of the Sun. *It is of stellar magnitude!* Objects in which gravity is balanced by the combined pressure of the gas and radiation must have a mass which is a few times the mass of the Sun.

A Limit on the Radiation Pressure

The inequality (4.9) also provides an *upper limit* to $(1 - \beta_c)$ for a star of a given mass M. Remember that $(1 - \beta_c)$ is just the ratio of the radiation pressure to the total pressure at the centre of the star. It follows from Eq. (4.9) that this upper limit is given by:

$$1 - \beta_c \le 1 - \beta_* \tag{4.12}$$

where $(1 - \beta_*)$ is uniquely determined by the mass M of the star and the mean molecular weight, μ, by the quartic equation:

$$\boxed{M = \frac{5.48}{\mu^2}\left[\frac{1-\beta_*}{\beta_*^4}\right]^{\frac{1}{2}} M_\odot.} \tag{4.13}$$

This quartic equation is simply obtained by rearranging Eq. (4.9) and using Eq. (4.11). Given a star of a certain mass, the above equation due to Chandrasekhar provides an *upper limit* to the radiation pressure at the centre of the star. In the case of the Sun, for example, *the radiation pressure at the centre cannot exceed 3 percent of the total pressure.*

Let us now summarize our discussion. We posed two questions at the beginning of this chapter concerning the *happening of the stars*:

(1) Why do the stars have nearly equal masses?
(2) Why is a *tussle* between gas pressure and radiation pressure important for the happening of the stars?

Eddington did not have a satisfactory answer to either of these questions. But Chandrasekhar answered both these questions in an authoritative and elegant manner.

1. Why do the stars have nearly the same mass?

Let us go back to Eddington's parable of the physicist in the cloud-bound planet. Recall that the physicist had never seen the sky. All he knew was laboratory physics, and natural constants measured in the laboratory. Based on his knowledge of physics, he constructs hypothetical objects in which the *inward pull due to gravity was balanced by the combined effect of gas pressure and radiation pressure.* His theory provides him with a *scale of mass* with which to measure such objects. And that mass scale, defined by a combination of fundamental constants occurring in the theory, is $29.2\,M_\odot$:

$$\left(\frac{hc}{G}\right)^{\frac{3}{2}} \frac{1}{m_p^2} \cong 29.2 M_\odot.$$

His theory thus tells him that the characteristic mass of such objects would be a few times 10^{33} g (recall that the mass of the Sun is 2×10^{33} g). They will not be of planetary mass, nor would they be thousands of times more massive than 10^{33} g.

This is the answer to the first question. Is it not extraordinary?

2. Why is a tussle between gas pressure and radiation pressure important for the happening of the stars?

Our physicist's theory also tells him that the *stability* of his objects requires that the ratio of radiation pressure to the total pressure at the centre must be less than a critical fraction dependent only on the mass of the star, as described in Eq. (4.13).

$$M = \frac{5.48}{\mu^2}\left[\frac{1-\beta_*}{\beta_*^4}\right]^{\frac{1}{2}} M_{\odot}.$$

This is the answer to the second question. One can now appreciate *why a* tussle *between gas pressure and radiation pressure is important for the happening of the stars.*

It turns out that nature *has* made objects similar to what our physicist friend had constructed in the laboratory. Not surprisingly, when Eddington draws aside the veil of clouds, our physicist finds that there are, indeed, thousands of millions of shining globes of gas in the sky, *the majority having masses between* 10^{33} *and* 10^{34} g.

What do we conclude from the foregoing discussion of *what are the stars and why are they as they are*? There is no better way to wind up this chapter than by quoting Chandrasekhar from his Nobel Prize Lecture (1984):

> We conclude that to the extent (4.13) is at the base of the equilibrium of actual stars, to that extent the combination of natural constants, $\left(\frac{hc}{G}\right)^{\frac{3}{2}} \frac{1}{m_p^2}$, providing a mass of proper magnitude for the measurement of stellar masses, is at the base of a physical theory of stellar structure.

Chapter 5
Energy Generation in the Stars

The Hypothesis of Nuclear Fusion in the Stars

In Chap. 2, we discussed the extraordinary conjecture by Eddington that the source of energy in the Sun and the stars *must be* the transmutation of the elements, more specifically, the conversion of hydrogen into helium. You will recall that he was led to this conclusion by the experimental findings of F. W. Aston, working in Rutherford's laboratory in Cambridge. Aston's interest was to measure the masses of atoms accurately. One of the discoveries he made was that *the mass of four hydrogen nuclei was greater than the mass of one helium nucleus*. Eddington's idea was that if four protons fuse to produce a helium nucleus, then this *mass deficit* would be converted into energy according to Einstein's formula:

$$E = \Delta M c^2.$$

Let us examine this closely. The mass of four protons is $4 \times 1.0081 m_u$ (atomic mass units), while the measured mass of the ^{4}He nucleus is $4.0039 m_u$. This means that a mass of $2.85 \times 10^{-2} m_u$ has disappeared for every helium nucleus produced if, indeed, the helium nucleus is produced by fusing four protons. This is roughly 0.7 percent of the original mass of hydrogen and corresponds to energy measuring about 26.5 MeV (Million electron-Volt, which is the unit usually used to measure energy in nuclear physics). Another way to say this is the following. If mass, M, of hydrogen is converted into helium, then the energy released is $0.007 M c^2$. (Think of James Bond to remember this formula!) The mass of the Sun is 2×10^{33} g, most of it hydrogen. By converting most of it to helium, it can generate $\sim 10^{52}$ erg of energy. The rate at which it radiates this energy (its luminosity) is 4×10^{33} erg s^{-1}. Therefore, the Sun can easily shine for 10^{11} years by tapping this source of *subatomic energy*.

G. Srinivasan, *What are the Stars?* Undergraduate Lecture Notes in Physics,
DOI: 10.1007/978-3-642-45302-1_5,

$$t_{\text{nuclear}} \sim \frac{0.007 M_{\odot} c^2}{L_{\odot}}.$$

$$t_{\text{nuclear}} \sim \frac{0.007 \times 2 \times 10^{33} \times 10^{21} \text{erg}}{4 \times 10^{33} \text{erg s}^{-1}} \sim 10^{11} \text{ years}. \quad (5.1)$$

This was Eddington's wonderful idea. In other words, the Sun generates energy by setting off *hydrogen bombs*—recall that in such a bomb hydrogen is fused into helium. The difference is in the scale! The energy released in a manmade bomb is roughly equivalent to converting a couple of kilograms of hydrogen into helium. The observed luminosity of the Sun implies that 600×10^{12} g of hydrogen is being converted into helium every second—that is, more than *600 million metric tons per second*! A natural question that comes to mind is, 'Why does the Sun not blow itself apart?' We shall return to this interesting question a little later.

The Basic Difficulty

While Eddington may have hit upon a brilliant idea, it is not at all obvious that four hydrogen nuclei can be *fused* to form a helium nucleus. The hydrogen nuclei (namely, the protons) are positively charged particles. Consequently, they will repel each other due to the electrostatic force

$$F_{\text{Coulomb}}(r) = \frac{e^2}{r^2}.$$

Given this repulsion, how can one bring two protons so close to one another that they practically touch? For only then will they bind together under the influence of the strongly attractive nuclear force. We need to answer to this question before we can understand how the transmutation of elements occurs. Let us, therefore, digress a little.

Interestingly, the basic clue regarding how nuclei might *fuse together*, despite their strong mutual repulsion, came from the phenomenon of natural radioactivity, namely, the spontaneous disintegration of heavy nuclei. Let us, therefore, discuss the problem of how *alpha particles* (^{4}He nuclei) are emitted by the uranium nucleus during its radioactive decay.

To understand why there is a problem in understanding this phenomenon, let us discuss the *inverse* of this, namely, the scattering of alpha particles by uranium nucleus (the famous scattering experiment by Rutherford). Since the alpha particle is positively charged, it will experience repulsion given by Coulomb's law:

$$V(r) = \frac{2Ze^2}{r}, \quad (5.2)$$

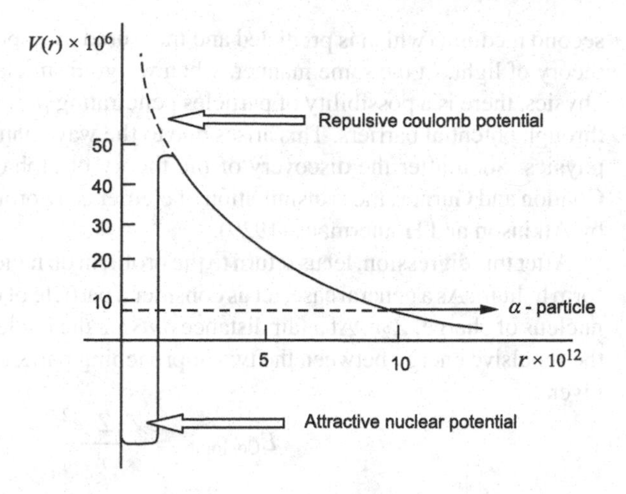

Fig. 5.1 A schematic diagram of the potential energy of an alpha particle particle in the field of a uranium nucleus

where Ze is the charge of the nucleus, and the alpha particle has two units of positive charge. The scattering experiments show that this repulsive force persists to a distance less than about 3×10^{-12} cm (see Fig. 5.1). But we know that at much smaller distances there must be a deviation from Coulomb's law since the stability of the uranium nucleus requires that there must be a *hole* in the potential energy at the centre of the nucleus. This hole arises due to the strong attractive nuclear force that dominates at very short distance $\sim 10^{-13}$ cm. So the potential energy curve looks like a high volcano with a very deep caldera. Yet the uranium nucleus emits alpha particles which have energy of 6.6×10^{-6} erg. It is very difficult to understand how the alpha particles trapped in the potential hole can escape. Since their kinetic energy is much smaller than the depth of the potential hole, surely, they cannot *climb over* the top of the well?

This great puzzle was solved in 1928 independently by the brilliant Russian physicist George Gamow, and by Condon and Gurney in the United States. The resolution of the problem invoked the newly emerging *quantum physics*. The underlying principle of quantum physics is the duality between particles and waves. It is this wave nature of particles that allows an alpha particle to escape from the nucleus. An analogy from optics (originally given by Gamow) will give us a feeling for how one may view this.

Imagine a beam of light incident on the boundary between two media at an angle greater than the *critical angle*. According to the laws of *geometrical optics,* we will have a total reflection of the incident beam—the light will be reflected at the interface between the two media and no disturbance occurs in the second medium. However, if the same problem is treated within the *wave theory of light,* it is found that there is, in fact, some disturbance in the second medium as well. This is the phenomenon of *evanescent waves*—a phenomenon which is appreciable for a distance of the order of a few wavelengths of light. The evanescent wave decays exponentially as we go into the second medium. There is no interpretation of this disturbance which occurs in the second medium (which is predicted and measured by experiment) in the geometrical theory of light. In the same manner, when we go from classical physics to quantum physics, there is a possibility of particles penetrating potential barriers, or *tunnelling* through potential barriers. This arises due to the wave nature of particles in quantum physics. Soon after the discovery of the theory of alpha decay by Gamow and by Condon and Gurney, the transmutation of elements by proton capture was considered by Atkinson and Houtermans (1929).

After this digression, let us return to the problem on hand, namely to fuse protons to form helium. As a general case, let us consider a particle of charge, $Z_1 e$, approaching a nucleus of charge, $Z_2 e$. At a fair distance *outside* the nucleus radius, $r_0(\approx 10^{-13}\,\text{cm})$, the repulsive energy between the two approaching particles due to Coulomb force is given by:

$$E_{\text{Coulomb}} = \frac{Z_1 Z_2 e^2}{r}. \tag{5.3}$$

Within the nucleus there is strong attraction between the particles. The depth of the potential hole is approximately 30 MeV. The superposition of the attractive nuclear force and the repulsive Coulomb force leads to a sharp potential jump. This is commonly referred to as the *Coulomb barrier*, and is typically of the order of:

$$E_{\text{Coulomb}}(r_0) \approx Z_1 Z_2 \text{ MeV}, \tag{5.4}$$

where Z_1 and Z_2 are the atomic number of the two colliding nuclei. For the case of two protons colliding against each other, the height of the Coulomb potential barrier is $\sim$1 MeV.

How does this energy compare with the typical energy of the colliding particles at the centre of the Sun? You will recall that the average temperature inside the Sun is $\sim$10 million kelvin, while the central temperature is $\sim$15 million kelvin. The *average energy* of the particles is approximately $k_B T$, where k_B is Boltzman's constant. It is often useful to express the energy in *electron volts.* For example, if the temperature is 10^4 K, then the average energy is 1 eV ($k_B \times 10^4$ kelvin $\approx$ 1 eV).

$$10^4 \text{ kelvin} \Leftrightarrow 1 \text{ eV}.$$

So at the central temperature of 10^7 K, the average energy is 1000 eV. This means that the typical energy of the protons is a *thousand times less than the height of the potential barrier,* which is $\sim$1 MeV. A proton with energy $\sim$1000 eV at infinity can never hope to climb the potential hill and fall into the hole at the centre. According to classical physics, it can only roll up the hill to a point where all its kinetic energy has been converted into potential energy (the point r_1 in Fig. 5.2); it is forbidden for the particle to temporarily *borrow* energy, climb up the hill and fall into the hole.

At this point we are tempted to feel that there is a way out. After all, $k_B T$ is only the *average energy* of the particles. Surely there must be particles with much greater energy than this; after all this is guaranteed to us by Maxwell's distribution of

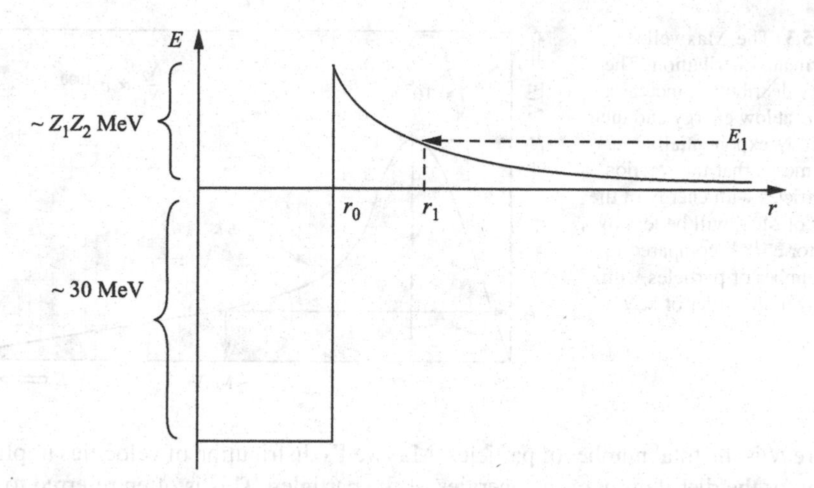

Fig. 5.2 The potential energy of a particle of charge, Z_1e, approaching a nucleus of charge, Z_2e: Coulomb repulsion dominates outside the nuclear radius, r_0. The height of the *Coulomb barrier* and the nuclear potential hole are indicated. In classical physics, the incident particle with energy E_1 will be reflected at the point r_1

velocities of particles. Let us recall what Maxwell taught us. Consider an enclosure of volume, V, containing gas at a temperature, T. By virtue of the heat energy, the particles of the gas will be in state of perpetual motion; some moving with greater velocity than others. An instant later, our particle will collide with another, changing its direction and moving with a different velocity. Such collisions—and they will be very frequent—will erase the *memory* of the particles. Therefore, one cannot make definitive statements about the velocity of any particular particle. One can only give a *statistical* or *probabilistic* description. In what must surely be one of the greatest discoveries in physics, James Clerk Maxwell discovered the statistical distribution of velocities of the particles in a box. He showed that the number of particles with velocities in the range $\vec{v}$ and $\vec{v}+d\vec{v}$ is given by

$$\boxed{N(\vec{v})d\vec{v} \propto e^{-\frac{mv^2}{2k_BT}}\, d\vec{v}}, \tag{5.5}$$

where the velocities range from $-\infty$ to $+\infty$. This is *Maxwell's distribution of velocities.* Let us look at the important features of this distribution. First, the distribution is symmetric in velocities about zero velocity. This is as it should be. If the number of particles moving with a certain *velocity* (speed and direction) is not precisely equal to the number moving in the opposite direction, then there will be a net velocity, and the box would start moving in that direction! Therefore the distribution of velocities has to be symmetric. The *constant of proportionality* in the above distribution is to be fixed by the condition:

$$\int_{-\infty}^{+\infty} N(v)d\vec{v} = N, \tag{5.6}$$

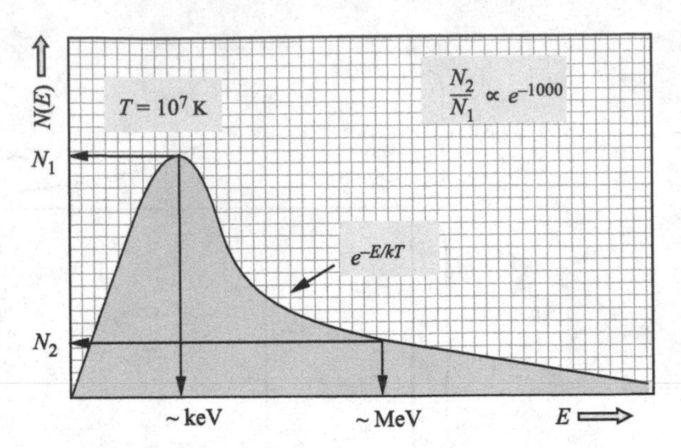

Fig. 5.3 The Maxwell–Boltzmann distribution. The energy distribution increases as $\sqrt{E}$ at low energy and then decreases exponentially (5.7). This means that the fraction of particles with energy of the order of MeV will be less by a factor e^{-1000} compared to the number of particles with energy of the order of keV

where N is the total number of particles. Maxwell's distribution of velocities implies a law for the distribution of the energies of the particles. This is often referred to as the *Boltzmann distribution*. Using the fact that $E = 1/2mv^2$, it is a simple matter to verify that Maxwell's law yields the following distribution for the number of particles having energy values in the interval E and $E + dE$:

$$\boxed{N(E)dE \propto e^{-E/kT}\sqrt{E}dE.} \tag{5.7}$$

The important thing to notice is that as one goes to larger energies, the fraction of particles decreases *exponentially* (Fig. 5.3).

Let us now return to our attempt to fuse two protons together by making them collide against each other. We concluded that since the average energy of protons (~1000 eV) is a thousand times less than the height of the Coulomb barrier (~MeV), the two protons can never come close enough to form a bound state. For this to happen, they have to come within the range of the attractive nuclear force. Our hopes had been raised by the recollection that even in a gas at a temperature of ten million kelvin, there will be particles with energy equal to or greater than the height of the potential barrier; perhaps these particles in the high-energy tail will do the job. Unfortunately, our hopes are misplaced. A look at the above distribution tells us that the fraction of particles with energy ~MeV will be less by a factor e^{-1000} compared to the number of particles with the average energy ~1 keV. This is an incredibly small number [$e^{-1000} \approx 10^{-434}$]. It is such a small fraction that even the fact that there are ~10^{57} protons in the Sun is not going to rescue us! The only possibility for fusion is due to the quantum *tunnel effect* discovered by Gamow.

Tunnelling Through a Potential Barrier

Lets us briefly discuss the penetration of particles through a potential barrier in quantum physics. We have already mentioned that the basic physics that underlies this possibility is the *wave–particle duality* in quantum physics. It is the *wave nature* of particles that makes the tunnelling probability nonzero.

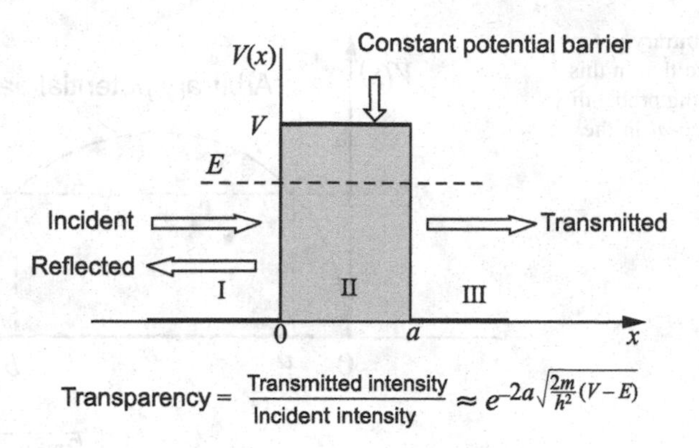

Fig. 5.4 Tunnelling through a constant potential barrier of height V. In classical physics, an incident particle with energy $E < V$ will be reflected by the barrier. In quantum physics, however, there is a finite probability for the particle to be transmitted through the barrier. This possibility arises because of the wave nature of particles in quantum physics. The tunnelling probability is, however, exponentially small

Let us first consider the example of a one-dimensional rectangular potential barrier, shown in Fig. 5.4. Consider a particle coming from the left with energy less than the height of the potential barrier. Classically, this particle cannot penetrate into region II and would be totally reflected. However, in quantum mechanics, since the incident particle is a *wave*, it will be partially reflected and partially transmitted (remember Gamow's analogy with evanescent wave in wave optics). The transparency or transmission probability can be calculated using wave mechanics (you will find the derivation in any introductory textbook). We shall merely state the result and highlight its main feature. Let us define the transparency or the tunnelling probability as

$$\text{Transparency} = \frac{\text{Transmitted intensity}}{\text{Incident intensity}}.$$

For a rectangular barrier of *height* V and width a,

$$\boxed{\text{Transparency} \approx \mathrm{e}^{-2a\sqrt{\frac{2m}{\hbar^2}(V-E)}}}. \tag{5.8}$$

Let us next consider a barrier of arbitrary shape as shown in Fig. 5.5.

In this case the transparency is given by

$$\boxed{\text{Transparency} \sim \mathrm{e}^{-2\int_a^b \sqrt{\frac{2m}{\hbar^2}(V(x)-E)}dx}}. \tag{5.9}$$

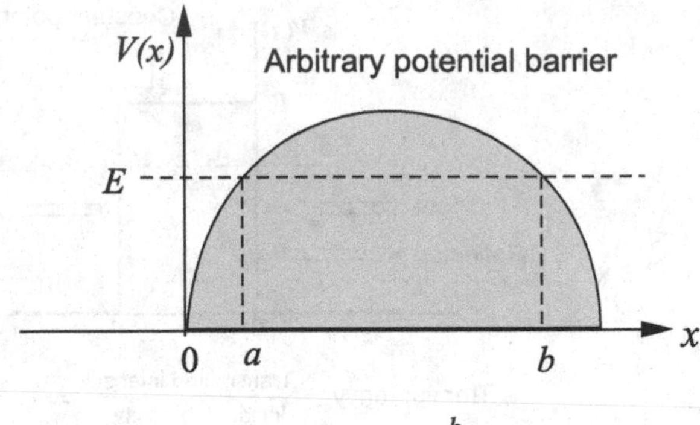

Fig. 5.5 An arbitrary potential barrier. Note that in this case the tunnelling probability involves an *integral* in the exponential

Look at the two expressions, (5.8) and (5.9), for the transmission probability. It is worth summarizing their most important features.

1. The tunnelling probability is an *exponential function.*
2. Given a barrier of a certain height V, the tunnelling probability *increases exponentially* with increasing energy of the incident particle.
3. The tunnelling probability *decreases exponentially* with increasing thickness of the barrier.
4. The probability is greater for particles of smaller mass.

After this brief review of quantum mechanical tunnelling, let us return to our problem. We are, of course, interested in a barrier arising because of the repulsive Coulomb potential. The solution of this problem can also be found in most of the standard books on quantum mechanics. The expression for the tunnelling probability through a Coulomb barrier is given below:

$$P \approx \mathrm{e}^{-2\pi G}. \tag{5.10}$$

The factor G is known as the Gamow exponent and is given by the expression

$$G \approx \sqrt{\frac{m}{2}} \frac{Z_1 Z_2 e^2}{\hbar E^{1/2}}. \tag{5.11}$$

Here m is the *reduced mass* of the nucleus and the colliding particle. We see that once again the tunnelling probability depends *exponentially* on the exponent G. That is, as G increases, the probability of tunnelling decreases exponentially. It will be seen from the approximate expression in (5.11) that *G increases with nuclear charge and decreasing energy of the particle.* This should be intuitively obvious because increasing Z_1 means that the particle will have to penetrate a *higher* potential barrier, and decreasing E implies that the particle has to tunnel through a *broader* potential barrier.

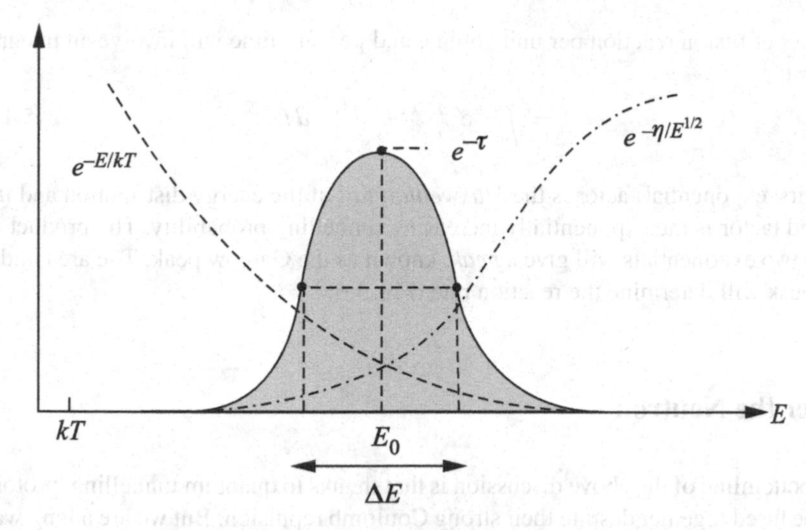

Fig. 5.6 The number of fusion reactions per unit volume per unit time is basically determined by the integral given in (5.12). The integrand is a product of two exponential functions. One of them is the tail of the Maxwell–Boltzmann distribution and it decreases exponentially (*dashed curve*). The other factor is the quantum mechanical tunnelling probability through a Coulomb barrier, and this rises exponentially with increasing energy (*dashed-dot curve*). Their product is a *bell-shaped curve* known as the *Gamow peak*. The hatched area under the Gamow peak determines the nuclear reaction rate. This is a schematic diagram, with each curve on a different scale

Let us now apply the above result for the collision of two protons ($Z_1 = Z_2 = 1$) with kinetic energy ~1000 eV (which is the average energy of the protons at a temperature of $T = 10^7$ K). If you plug in the values for the various constants that appear in the formula, you will find that the tunnelling probability, P, is of the order of 10^{-20} for particles with the average energy. You might feel a bit disappointed that the probability is still only ~10^{-20}. But this is a substantial probability, given that we have 10^{57} protons colliding against each other. It is like scoring a goal in a football match—even if the probability of the ball going into the net is very small, if you kick towards the goal sufficient number of times, you will score goals!

We are now in business, so to speak. Assisted by quantum mechanical tunnelling, we can hope to fuse protons together to form helium, thereby releasing vast amount of energy. Although the majority of protons will have energy equal to the *average value*, one should not forget the small number of particles with much larger energy. They play an important role in quantum tunnelling. It is true that the *fraction of particles decreases exponentially* with increasing energy (recall our discussion of the Maxwell–Boltzman distribution). But we just saw that the *tunnelling probability increases exponentially* with increasing energy of the particles. So our success rate in fusing two protons together (the *reaction rate*, in the technical jargon) will depend upon an interplay between two opposite trends: *an exponentially increasing tunnelling probability with increasing energy* and *an exponentially decreasing fraction of particles with increasing energy*. A more careful analysis will show that the number of fusion reaction per unit volume and per unit time will involve an integral of the type:

$$J = \int_0^\infty e^{-E/kT} e^{-\eta/\sqrt{E}} dE. \tag{5.12}$$

The first exponential factor is the *Maxwellian tail* of the energy distribution and the second factor is the exponentially increasing tunnelling probability. The product of these two exponentials will give a *peak*, known as the Gamow peak. The area under this peak will determine the reaction rate (Fig. 5.6).

Enter the Neutron

The bottomline of the above discussion is that thanks to quantum tunnelling, protons can be fused together despite their strong Coulomb repulsion. But we are a long way from filling in the details of how to form a helium nucleus starting with four protons. There is another fundamental problem to be overcome. While the mass of the helium nucleus is roughly equal to the combined mass of four protons—suggesting that somehow the fusion of four protons might have resulted in a helium nucleus—the helium nucleus has only two units of positive charge and not four. Although this particular difficulty was not bothering physicists—since they were not preoccupied with energy production in stars—they were concerned with a related problem.

Lord Rutherford was very concerned with the general difficulty in reconciling Bohr's theory of electrons rotating around the proton-filled nucleus and the *isotopes of elements*. According to Bohr's model of the atoms, the number of positively charged particles inside the nucleus must be equal to the number of orbiting electrons. This means that once we specify the atomic charge, the atomic mass (which is essentially the mass of the nucleus) should be uniquely determined. But Rutherford and his colleagues had discovered that many elements had isotopes. *All the isotopes of a given element had the same atomic charge but different atomic mass*. This meant that while the charge of the nucleus of all the isotopes of an element was determined by the number of electrons, the mass of the nucleus varied. This is only possible if the nucleus contained *neutral particles* in addition to the protons. If the number of these neutral particles was different in the different isotopes, then that would explain why their nuclear masses were different. Since no neutral particle was known in 1930, Rutherford was forced to postulate the existence of a *neutral doublet*—a bound pair made up of an electron and a proton—although there was no experimental proof of this. This is exactly what Eddington did to produce a helium nucleus out of four protons. He packed two electrons into the nucleus to make the net positive charge of the helium nucleus equal to two!

James Chadwick was to unlock the door to this basic problem in 1932. Chadwick was Rutherford's student at the University of Manchester and moved with his master to the Cavendish Laboratory in Cambridge in 1919. He became Rutherford's trusted assistant during the period of intense creativity at the Cavendish Laboratory. In 1932,

Joliot and Curie published the observation that alpha particles incident on beryllium produced evidence of carbon and an intense, 55-MeV gamma ray. Chadwick and Rutherford immediately realized that this result must be wrong; the energy of the gamma ray was too high and they suspected that a *neutral particle* must be involved. Within weeks, Chadwick established the reaction:

$$^{9}\mathrm{Be} + {}^{4}\mathrm{He} \rightarrow {}^{12}\mathrm{C} + {}^{1}\mathrm{n}$$

The particle on the right-hand side was a *neutral* particle, and was christened the *neutron.* Chadwick determined the ratio of the mass of the neutron to that of the proton to be 1.0090 (the current. This neutral particle could penetrate even lead. Chadwick received the Nobel Prize for this discovery in 1935. With this discovery, the list of *elementary particles* had grown to three: the electron, proton and the neutron. One was now in a position to explain the *isotopes* of the elements. *Isotopes of an element have the same number of protons but differ in the number of neutrons in the nucleus.*

The Neutrino

Even as this mystery was unfolding, there was another major puzzle. This concerned the *radioactive decay* of some of the elements. The key player in this drama is so central to the rest of our discussion that we shall narrate this story.

You will remember that in 1896 the French scientist, Becquerel, discovered the phenomenon of *β decay* in which some elements spontaneously emitted electrons, which were known as β rays at that time. Soon it was discovered that the β rays could be positively charged as well. Today we know that these are the positrons. The most remarkable aspect of this phenomenon was the apparent failure of energy conservation. In α decay, for example, energy was clearly conserved. If a nucleus A with energy E_A decays to nucleus B with energy, E_B, emitting an α particle, then

$$E_A = E_B + E_\alpha,$$

where E_α is the energy of the α particle. In the case of β decay, such an equation could not be written. The reason is that in the decay of any given nucleus the energy of the emitted electrons is a *continuous spectrum*, as shown in Fig. 5.7.

This means that one *cannot* write an equation like $E_A = E_B + E_\beta$ since the energy of the electron can have any value less than the maximum. This is why many physicists, including the great Neils Bohr, were willing to entertain the notion that energy conservation may be violated in β decay. But Wolfgang Pauli would have none of this. He tried to rescue energy conservation by postulating that another particle was emitted in β decay. His idea was that this additional particle would carry away the missing energy. Pauli was very clever. He made sure that this mysterious particle is very difficult to detect! He did this by saying that it must not only be electrically

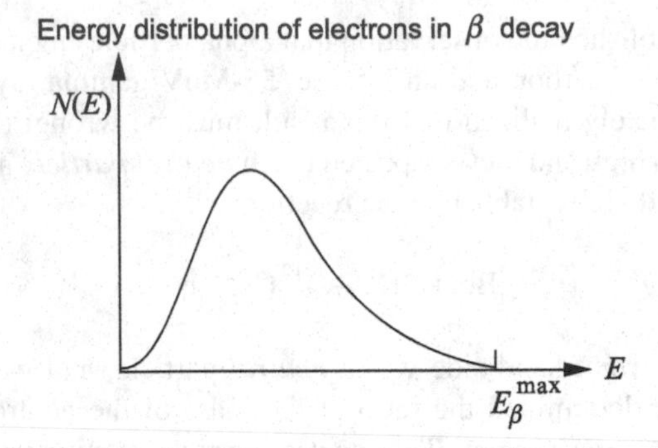

Fig. 5.7 The most intriguing thing about β decay was that the electrons emitted in radioactive decay had a continuous spectrum, with a high-energy cut off. This seemed to defy the sacred principle of *energy conservation* because one could not write an equation of the type $E_A - E_B = E_\beta$, where E_A and E_B are the energies of the parent and daughter nuclei, respectively. This great puzzle led Wolfgang Pauli to postulate the existence of a neutral particle emitted along with the electron, which Enrico Fermi named the *neutrino*

neutral, but its mass must be very nearly zero. Pauli did not advance this radical suggestion in a scientific publication. He merely mentioned it in a letter which he wrote in 1930:

> I have come upon a desperate way out regarding the ⋯ continuous β-spectrum ⋯. There could exist in the nucleus elementary neutral particles, which I shall call neutrons ⋯
>
> I admit that my way out may not seem very probable a priori since one would probably have seen this neutron a long time ago if they exist. But only one who dares wins⋯. One must therefore discuss seriously every road to salvation⋯.

As we discussed in the "Enter the Neutron", a *neutral particle* was discovered by Chadwick in 1932. But that could not possibly be the particle postulated by Pauli because it was a *heavy* particle with a mass almost equal to the mass of the proton. Nevertheless, Chadwick had christened it the *neutron.*

The next major development in β decay took place in 1933 when the great Italian physicist Enrico Fermi advanced a major hypothesis that radioactivity arises due to a new kind of force, which has come to be known as the *weak force.* The list of fundamental forces in nature had grown to four: *electromagnetic force*, *gravitational force*, *strong force* (which held the constituents of the nucleus together) and *weak force*. He went on to construct a theory of this weak force in analogy with the *electromagnetic force.* Fermi's identification of a new kind of force, and constructing a theory of *weak interaction* mediated by this force, was undoubtedly one of the great intellectual achievements of the twentieth century. If you are interested in learning more about this, and how the exciting story of elementary particles unfolded in the twentieth century, I would recommend that you look at *The Big and the Small* by G. Venkataraman.

According to Fermi, radioactivity should be understood in terms of the following two basic processes that occur inside the nucleus:

Table 5.1 Members of the lepton family

Name	Symbol	Lepton number
Electron	e^-	+1
Electron neutrino	ν_e	+1
Electron antineutrino	$\overline{\nu}_e$	−1
Muon	μ	+1
Muon neutrino	ν_μ	+1
Muon antineutrino	$\overline{\nu}_\mu$	−1
Tau	τ	+1
Tau neutrino	ν_τ	+1
Tau antineutrino	$\overline{\nu}_\tau$	−1

$$\begin{array}{l} \text{neutron} \rightarrow \text{proton} + \text{electron} + \text{neutrino} \ \ (n \rightarrow p + e^- + \nu) \\ \text{proton} \rightarrow \text{neutron} + \text{positron} + \text{neutrino} \ \ (p \rightarrow n + e^+ + \nu) \end{array} \tag{5.13}$$

You will notice that Fermi had invoked the neutral particle postulated by Pauli and christened it the *neutrino*—the *little neutron* (in Italian).

Some years later, physicists introduced the concept of the *antiparticle* of the neutrino, namely the *antineutrino,* usually denoted by $\bar{\nu}$. This was in analogy with the antiparticle of the electron, namely the *positron*, predicted by Dirac's relativistic theory of the electrons.

Meanwhile, the family of *leptons* (the light particles) grew, with the *mu meson* (μ) and the *tau meson* (τ) added to the list. It then turned out that there was a neutrino and an antineutrino associated with each of these new arrivals to the lepton family. Thus, there are *six* members of the neutrino family (Pauli would have been amused!). The symbols and the associated quantum number, namely the lepton number, are given in Table 5.1.

With this modern perspective, let us go back to the original suggestion made by Fermi and write the β decay Eq. (5.13) properly:

$$\begin{array}{l} \text{neutron} \rightarrow \text{proton} + \text{electron} + \text{antineutrino} \ (n \rightarrow p + e^- + \bar{\nu}_e) \\ \text{proton} \rightarrow \text{neutron} + \text{positron} + \text{neutrino} \ (p \rightarrow n + e^+ + \nu_e) \end{array} \tag{5.14}$$

In rewriting Fermi's equations, we have been careful to indicate that it is the *electron neutrino* and its antiparticle that enter the two decays described by Eq. (5.14). Written in this form, the above two decays satisfy several conservation laws:

1. Conservation of *energy*
2. Conservation of *charge*
3. Conservation of *baryon number* [both the proton and neutron have a baryon number of +1]

4. Conservation of *lepton number* [+1 for the electron and -1 for the positron or antielectron; +1 for the neutrino and -1 for the antineutrino]

Where did the electron and neutrino come from? Were they inside the nucleus? Fermi said that they were both spontaneously created at the time of the decay. Remember that it is the same in the case of photons emitted by atoms. Photons do not exist inside the atom. They are created spontaneously.

Well, this is how the elusive neutrino entered the scene. Pauli was right in saying that it must interact very weakly with matter. Indeed, it took more than twenty years to find the neutrino. Reines and Cowan designed a careful experiment to detect the antineutrinos emitted by neutrons (refer to Eq. 5.14) in a nuclear reactor and struggled for many years. Their patience and perseverance was finally rewarded in 1956, twenty-six years after Pauli had postulated the existence of the neutrino. Astonishingly, it took forty more years for the Nobel Committee to recognize this extremely important discovery! This is particularly surprising since in the intervening decades several Nobel Prizes were awarded for the discovery of new fundamental particles. Unfortunately, by the time the Nobel Prize was awarded in 1995, Cowan was no more.

Why did it take more than twenty years to discover the neutrino? The reason is very simple—they interact very weakly with matter. Way back in 1938, Hans Bethe and Rudolf Peierls estimated the scattering cross-section for neutrinos and concluded that neutrinos would travel many *light years* before they interacted with an atom. The scattering cross-section σ_ν for interaction with matter increases with increasing energy. A rough estimate gives:

$$\sigma_\nu \approx (E_\nu / m_e c^2)^2 10^{-44}\ \mathrm{cm}^2. \tag{5.15}$$

In the above formula, the energy of the neutrino is expressed in units of the rest mass energy of the electron; the moral is that low-energy neutrinos have a smaller cross-section and would therefore be more difficult to detect. Neutrinos with energy in the MeV range would have a cross-section $\approx 10^{-44}\mathrm{cm}^2$, *which is smaller than the typical cross-section for photons in matter by a factor* 10^{-18}. Sometimes it is easier to think in terms of the *mean free path* rather than the cross-section. Recall that this is the mean distance travelled by the particle between two successive collisions (we encountered this concept while discussing the opacity of stellar matter). The mean free path of the neutrinos is given by the approximate expression

$$l_\nu \approx \frac{2 \times 10^{20}\mathrm{cm}}{\rho} \tag{5.16}$$

where ρ is the density of matter in cgs units. At the stellar density of 1 g cm^{-3} the mean free path of the neutrinos is more than 100 *light years*!

The Synthesis of Helium in the Stars

Now we are all set to discuss some details of the energy generation in the stars. The first breakthrough in solving the problem of how stars liberate energy came in 1938 when C. F. von Weizsäcker discovered a nuclear cycle, now known as the *carbon–nitrogen–oxygen (CNO) cycle*, in which hydrogen nuclei could be fused using carbon as a catalyst. However, von Weizsäcker did not work out the rate at which energy could be produced in the stars using this CNO cycle or how this rate would depend on the temperature that obtains in the stars.

The credit for this must go to Hans Bethe, the acknowledged master of nuclear physics. In 1938, Bethe had just completed a set of three monumental review articles in nuclear physics. These were known as *Bethe's Bible.* The first textbooks in nuclear physics were published only several years after the end of World War II. Until then, physicists all over the world learnt their nuclear physics from these pedagogical and authoritative articles by Bethe. We have already remarked that in the 1930s physicists were not concerned with problems in astronomy. They were more interested in atomic and molecular spectra and nuclear physics. It was George Gamow who sensitized physicists about the unsolved problems concerning stellar physics by convening a small conference in Washington, D. C. Hans Bethe and many of the leading physicists were at that conference. Within a few months of this, Hans Bethe had worked out, in great detail, the synthesis of helium in stars and published his results in a landmark paper entitled, 'Energy Production in Stars' (1939). Bethe considered two processes. One of them has come to be known as the *p–p chain* in which one builds helium out of hydrogen. This is the process that is important for stars like the Sun, and stars of even lower mass. The other process is the *CNO cycle* discovered earlier by von Weizsäcker, and is the dominant process for stars more massive than the Sun.

The p–p Chain Reaction

Let us first discuss the p–p chain reaction. This reaction derives its name from the first reaction between two protons forming a deuterium nucleus.

$$\boxed{{}^{1}\mathrm{H}+{}^{1}\mathrm{H} \rightarrow {}^{2}\mathrm{H}+\mathrm{e}^{+}+\nu.} \tag{5.17}$$

The *deuterium* nucleus (^{2}H) has one proton and one neutron. Since the right-hand side also must have two positive charges, a positron or *antielectron* (e^+) is also produced. The last particle on the right-hand side is the *neutrino.* Clearly, for this reaction to occur a proton must have transformed itself into a neutron:

$$\mathrm{p} \rightarrow \mathrm{n}+\mathrm{e}^{+}+\nu.$$

But a free proton cannot spontaneously decay into a neutron. You may say, 'But Fermi said so!' Fermi was careful to consider the decay of the neutron and a proton *inside* the nucleus. In 1933, there was no evidence for the decay of even a free neutron. The reason why a free proton cannot decay into a neutron is that a proton is lighter than a neutron, and so it would violate the conservation of energy. However, a proton *in close proximity of another proton can decay;* it can borrow energy and momentum from the passerby. But this reaction is governed by *weak interaction,* and the rate is too slow to be measured in the laboratory. However, the rate can be calculated accurately using the theory of low-energy weak interaction, and the calculated lifetime for the reaction (5.17) to occur under the conditions that obtain in the solar interior is about *1010 years*! Given this, it would seem ridiculous to entertain the reaction indicated in (5.17). But we must remember that we are dealing with 1057 protons colliding with each other. So the formation of deuterium is not so improbable. It happens in the stars. And if it did not happen, the Sun would not be shining!

Once a deuterium nucleus forms, the next reaction proceeds very fast (in a fraction of a second).

$$\boxed{{}^{2}\mathrm{H} + {}^{1}\mathrm{H} \rightarrow {}^{3}\mathrm{He} + \gamma} \tag{5.18}$$

The formation of ^{4}He now proceeds via one of three alternate branches. (See Figs. 5.8 and 5.9 for illustrations of the synthesis and decay of ^{4}He.)

Branch 1 (85 percent)

$$\boxed{{}^{3}\mathrm{He} + {}^{3}\mathrm{He} \rightarrow {}^{4}\mathrm{He} + {}^{1}\mathrm{H} + {}^{1}\mathrm{H}} \tag{5.19}$$

Branch 2 (15 percent)

$$\boxed{\begin{gathered} {}^{3}\mathrm{He} + {}^{4}\mathrm{He} \rightarrow {}^{7}\mathrm{Be} + \gamma \\ {}^{7}\mathrm{Be} + \mathrm{e}^{-} \rightarrow {}^{7}\mathrm{Li} + \nu \\ {}^{7}\mathrm{Li} + {}^{1}\mathrm{H} \rightarrow {}^{4}\mathrm{He} + {}^{4}\mathrm{He} \end{gathered}} \tag{5.20}$$

Branch 3 (0.01 percent)

$$\boxed{\begin{gathered} {}^{3}\mathrm{He} + {}^{4}\mathrm{He} \rightarrow {}^{7}\mathrm{Be} + \gamma \\ {}^{7}\mathrm{Be} + {}^{1}\mathrm{H} \rightarrow {}^{8}\mathrm{B} + \gamma \\ {}^{8}\mathrm{B} \rightarrow {}^{8}\mathrm{Be} + \mathrm{e}^{+} + \nu \\ {}^{8}\mathrm{Be} \rightarrow {}^{4}\mathrm{He} + {}^{4}\mathrm{He} \end{gathered}} \tag{5.21}$$

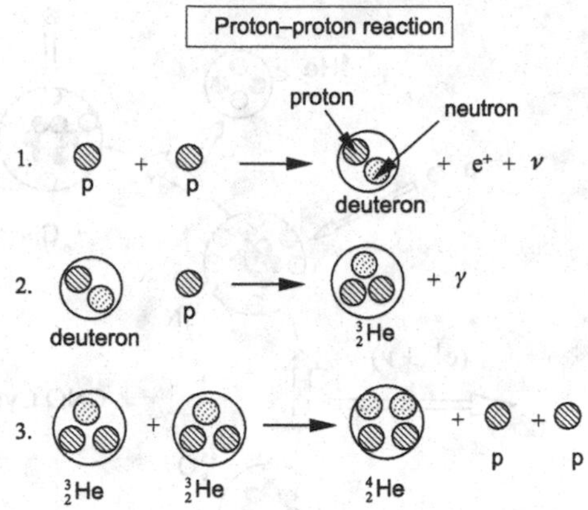

Fig. 5.8 The synthesis of protons into helium nucleus. This is *Branch 1* of the p–p chain reaction and accounts for 85 percent. The remaining 15 percent is through *Branches 2* and *3*, explained in (5.20) and (5.21)

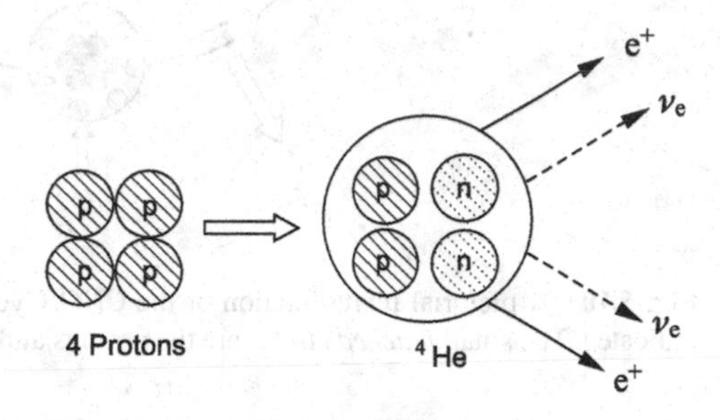

Fig. 5.9 In both the p–p reaction chain and the CNO cycle, four protons are involved in forming a helium nucleus. For every helium nucleus that is synthesized, *two positrons* and *two electron neutrinos* are emitted

The *percentages* given in parentheses refer to the relative importance of the three branches. *The end result of all these reactions is to fuse four protons together to form a helium nucleus, with the release of two positrons* (e^+), *two neutrinos* (ν) *and about 26.7 MeV of energy in the form of gamma rays and kinetic energy of created particles.*

The CNO Cycle

The other route for the synthesis of helium is the carbon–nitrogen–oxygen cycle, first discovered by C. F. von Weizsäcker. The details of this were worked out by Hans Bethe in 1939. The CNO cycle requires the presence of some carbon, nitrogen or oxygen which act as catalysts in chemical reactions. Here also, four protons are fused into one helium nucleus, releasing roughly the same amount of energy as before (25 MeV per ^{4}He nucleus produced).

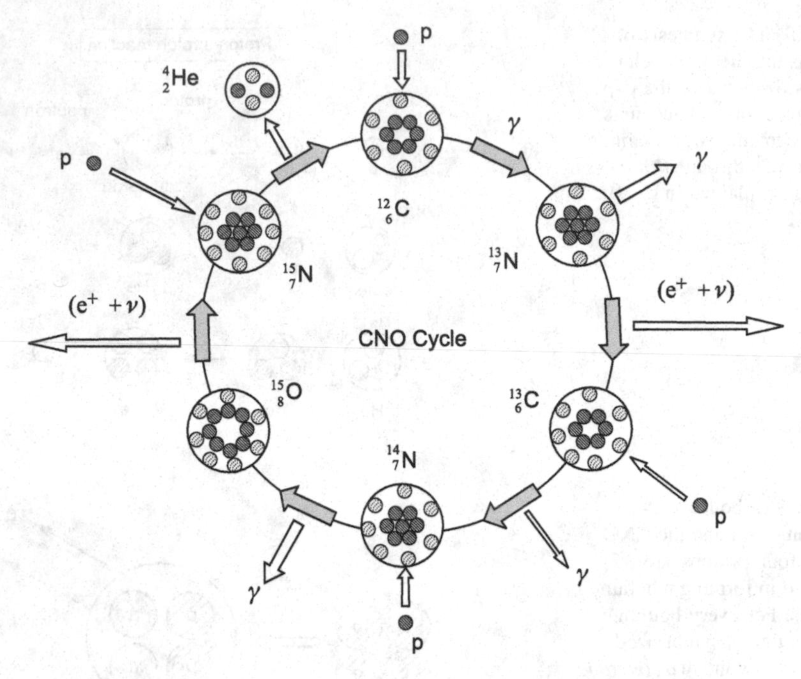

Fig. 5.10 A pictorial representation of the CNO Cycle. The *bigger circles* represent the nuclei indicated. The small *hatched circles* are the protons and the *small circles* with *dots* are the neutrons

$$\begin{array}{|c|}\hline {}^{12}C + {}^{1}H \rightarrow {}^{13}N + \gamma \\ {}^{13}N \rightarrow {}^{13}C + e^{+} + \nu \\ {}^{13}C + {}^{1}H \rightarrow {}^{14}N + \gamma \\ {}^{14}N + {}^{1}H \rightarrow {}^{15}O + \gamma \\ {}^{15}O \rightarrow {}^{15}N + e^{+} + \nu \\ {}^{15}N + {}^{1}H \rightarrow {}^{12}C + {}^{4}He \\ \hline \end{array} \tag{5.22}$$

Notice that initial ^{12}C nucleus is recovered after a cycle of reactions. In this cycle also two positrons and two neutrinos are emitted. The CNO cycle is the dominant mechanism for the synthesis of helium in stars more massive than the sun. As already mentioned, the p–p chain reaction is more important for the Sun, with the CNO cycle contributing only about 1.5 percent to the energy generation (Fig. 5.10).

The remarkable thing is that Hans Bethe was not just content with outlining these ideas in his monumental paper. He authoritatively analysed the different possibilities and selected as the most important among these the two processes we have outlined above. The hallmark of Bethe's scientific papers was the thoroughness with which every detail is attended to, and this paper was no exception. Hans Bethe was undoubtedly one of the greatest physicists of the twentieth century. The brief biographical sketch given on the following page will give you some idea about this great man.

Why Does the Sun Not Blow Itself Up?

Before proceeding, let us discuss the interesting question we had raised earlier. The proton–proton chain reaction we have just discussed is the same as the reaction which takes place in hydrogen bombs. We mentioned at the beginning of this chapter that the observed luminosity of the Sun implies *four million metric tons of hydrogen being converted to helium every second*. So why does the Sun not blow itself up? The fact that the Sun has been shining steadily for billions of years implies that a *safety valve* must be in operation. How does this safety valve work?

Hans Albrecht Bethe
1906–2005

Hans Albrecht Bethe was born in Strasbourg, Alsace-Lorraine, on 2 July 1906. After finishing his high school and early years in college, he went to do his Ph.D. with Professor Arnold Sommerfeld in Munich. At that time, Sommerfeld's group consisted of an impressive number of truly outstanding young persons such as *Wolfgang Pauli, Werner Heisenberg, Gregor Wenzel* and many others. These young men went on to create the new quantum physics under the watchful eyes of their mentor.

Bethe emigrated to England in 1933 after the rise of Adolf Hitler and the Nazis in Germany. After a year at the University of Manchester, and another year at the University of Bristol, he moved to Cornell University, USA in 1935. He remained there until his death in 2005!

Bethe was a true pupil of Sommerfeld. He was a master of theoretical physics and had a great understanding of experimental physics. He was influenced by Fermi's simplicity and Sommerfeld's rigor in approaching problems, and these qualities influenced his own later research. Thoroughness and scholarship characterized his work throughout a remarkably long career lasting nearly eighty years (Fig. 5.10).

Like many of the great physicists of that era, his research spanned all branches of physics. At some stage or the other, he made seminal contributions to atomic physics, solid state physics, statistical physics, nuclear physics and astrophysics.

Bethe's major contributions were in the theory of atomic nuclei. During the period he spent in England, he developed a theory of the deuteron together with Rudolf Peierls (another brilliant German physicist who had taken shelter in England). At Cornell he concentrated on the theory of nuclear reactions, predicting many reaction cross-sections. In connection with this work, he developed Bohr's theory of the compound nucleus in a more quantitative fashion.

This work, and also the existing knowledge on nuclear theory and experimental results, was summarized in three classic articles in the *Reviews of Modern Physics* which he wrote together with Livingston and Bacher, two of his young colleagues at Cornell. For many years these three articles served as a textbook for nuclear physicists and came to be known as 'Bethe's Bible.' It was during this period that he made his monumental contributions to stellar nucleosynthesis and energy production in the stars. Although this work earned him the *Nobel Prize for Physics*, it came only in 1967—nearly thirty years after his papers were published.

In 1947, Bethe was the first to explain the *Lamb shift* in the hydrogen spectrum, and he thus laid the foundation for the modern development of quantum electrodynamics. This work provided the impetus for the later work done by Richard Feynman, Julian Schwinger and Tomanaga which marked the beginning of modern quantum electrodynamics and for which they were awarded the Nobel Prize in 1965.

The Manhattan Project and the atom bomb

Bethe was a key player in the design and building of the first atom bomb. During the summer of 1942 he participated in a special session at the invitation of *Robert Oppenheimer*, which outlined the first designs for the atomic bomb. Initially, Bethe was skeptical of the possibility of making a nuclear weapon from uranium. In the late 1930s, he wrote a theoretical paper arguing against fission, but was convinced by Edward Teller to join the Manhattan Project. When Oppenheimer was put in charge of forming a secret weapons design laboratory at Los Alamos, he appointed Bethe the Director of the Theoretical Division, a move that irked Teller, who had coveted the job for himself.

Bethe's work at Los Alamos included calculating the critical mass of uranium-235 and the multiplication of nuclear fission in an exploding atomic bomb. Along with Richard Feynman, he developed a formula for calculating the explosive yield of the bomb. After November 1943, when the laboratory had been reoriented to solve the implosion problem of the plutonium bomb, Bethe spent much of his time studying the hydrodynamic aspects of implosion, a job which he continued into 1944. In 1945, he worked on the neutron initiator, and later on radiation propagation from an exploding atomic bomb.

After the end of World War II, Bethe played a key role in the development of the *hydrogen bomb.* Although initially he was very much against the development of this weapon, and hoped that it would never work, he decided to join the effort after the outbreak of the Korean War.

As we shall see in Chap. 7 of this monograph, when he was about eighty years old, he wrote an important article about the solar neutrino problem in which he dealt with the conversion of electron neutrinos into muon neutrinos. This idea was proposed to explain the discrepancy between theory and experiment. Bethe continued to do research on supernovae, neutron stars, black holes and other problems in theoretical astrophysics into his late nineties!

Hans Bethe died on 6 March 2005.

Let us recall our earlier discussion of the hydrostatic equilibrium of the stars. Our basic premise was that the inwardly directed force of gravity is balanced by the pressure of the gas. Let us pose our question once again. The synthesis of helium releases ~27 MeV of energy per nucleus. This energy will soon be converted into *heat.* What this really means is that the energy released goes into increasing the energy in the random motions of the particles; this is what one means by *heat.* This heat energy diffuses out, being carried by the radiation.

Let us say that there is some fluctuation and that some extra energy is liberated by the nuclear reactions. The central temperature and consequently the energy of the protons will increase. Now refer to Eqs. (5.10) and (5.11). The tunnelling probability increases exponentially with increasing energy. Hence, the rate of fusion reaction and therefore the rate of energy production depends very sensitively on the central temperature; the reaction rate will increase dramatically with an increase in the central temperature. This, in turn, will result in the heat energy increasing even more. We shall therefore have a *positive feedback.* At some stage, the heat energy will become so large that it will overwhelm the gravitational binding energy and the star will explode.

That this does not happen is due to the fact that the stellar material behaves as an *ideal gas.* An important property of an ideal gas is that its pressure is determined by its temperature. So, if the central temperature goes up in response to increased energy generation, then the gas pressure will increase (recall that for an ideal gas $P = nk_BT$). This increase in pressure will upset the hydrostatic equilibrium of the core of the star, and the core will consequently expand, lifting the overlying layers. The energy for this *lifting* will come at the expense of the internal energy of the

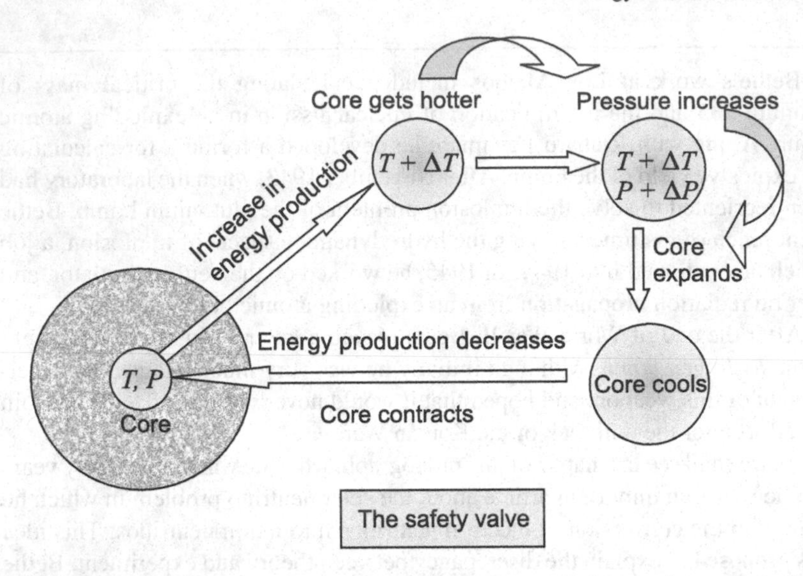

Fig. 5.11 The safety valve that prevents the Sun from blowing up. An increase in energy production would lead to the core getting hotter. Because the stellar material is an ideal gas, an increase in the central temperature will lead to an increase in the pressure. This, in turn, will lead to the expansion of the core, with a consequent decrease in the temperature and pressure. This will result in the core contracting to its original radius, temperature and pressure. This is how the safety valve works

gas. (Recall that according to the First Law of Thermodynamics, the energy spent in doing *work* comes at the expense of the internal energy). The core will cool as a result of this expansion and the rate of energy production will decrease, leading to an equilibrium situation. This is how the safety valve works (see Fig. 5.11).

But it is important to appreciate that such a safety valve is not guaranteed for all stars (see Fig. 5.12). There are situations when the pressure of a gas does *not* respond to a change in the temperature. For example, it could be that the pressure is *independent* of temperature. Then we shall certainly have a positive feedback, and a stellar explosion! We shall have occasion to discuss this in the next volume.

Where is the Evidence?

Eddington must have been overjoyed to read Bethe's paper. Bethe had not only worked out the details, he had got the answer that Eddington had predicted nearly twenty years earlier. Just to convince himself, Bethe used his results to work backwards to estimate the central temperature of the sun. Astonishingly, he got a value very close to what Eddington had earlier estimated (about 15 million degrees kelvin). This was a truly great triumph for theoretical physics. But can all this be just a coincidence? Is there any way to directly test the hypothesis that the Sun and the stars shine because deep in their interior they are converting hydrogen to helium. Unfortunately,

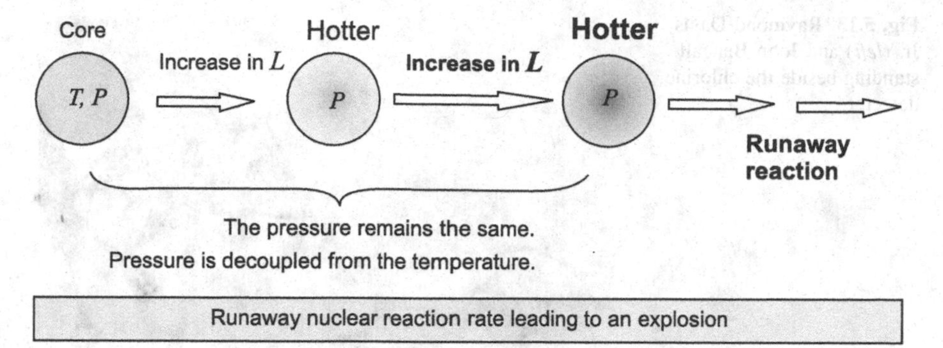

Fig. 5.12 The safety valve described in Fig. 5.11 is not guaranteed to work under all circumstances. As we shall discuss in the next volume, if the stellar material obeys the rules of quantum physics, rather than classical physics, then gas pressure is independent of temperature. If such a situation prevails, then an increase in the central temperature will result in a positive feedback and thereby a runaway nuclear reaction. Sometimes, this could result in an explosion of the star

we cannot *see* the interior. The light that we receive originates as high-energy radiation near the centre. This radiation takes several million years to diffuse out to the surface, being degraded in energy in the process. Finally, when this radiation emerges as *visible radiation,* it tells us only about the conditions that prevail near the surface. So the radiation that finally escapes does not tell us anything about what is happening near the centre; it merely tells us about the conditions near the surface.

That leaves us with two other potential witnesses, the *positrons* and the *neutrinos* which are released in the fusion reactions. The positron (which is the antiparticle of the electron) will not get very far from its birthplace since it will soon encounter a free electron, and the electron–positron pair will get annihilated. Remember that when a particle and its antiparticle collide, their rest mass energy is converted to two or more photons [It cannot be just a single photon. Think about why this is so!]. That leaves us with the neutrinos as the only other witnesses to have been at the scene of the crime. But the neutrinos are *hostile witnesses* for they do not like to be questioned. They quickly leave the scene of the crime. We have already mentioned that neutrinos interact so weakly with matter that their mean free path is many light years at terrestrial densities, let alone the density that obtain inside the Sun. Consequently, the neutrinos produced in the fusion reactions will escape from the Sun. Unfortunately, they will pass through the Earth also. The hope is that since *the Sun emits more than* 10^{38} *neutrinos per second* there is a finite probability that a few of them will interact with terrestrial matter—if we wait long enough. Despite this great difficulty, if we could detect the neutrinos from the Sun and if their flux and energy agree with theoretical predictions, then it would be a direct confirmation of Eddington's conjecture that the Sun is, in fact, converting hydrogen to helium at its centre.

In 1964, a well-known experimentalist Raymond Davis Jr. and an equally well-known young theoretical physicist John Bahcall (shown in Fig. 5.13) proposed an experiment to detect the neutrinos from the Sun. Four years later, Davis succeeded

Fig. 5.13 Raymond Davis Jr. (*left*) and John Bahcall standing beside the chlorine detector

in detecting the solar neutrinos. And on that note, I could end this book. Instead, I shall devote two more chapters(!) to a discussion of the results of this and other experiments, as well as some related developments, for this will exemplify how science progresses. We have read, and have been taught, about the great discoveries by Tycho Brahe, Johannes Kepler and others. But we are seldom told how painstaking the process of making great discoveries tends to be. Often, there is a tendency to romanticize science. The story of the solar neutrinos provides an excellent illustration of what a real grind science is. And the story of the solar neutrinos I am about to narrate beats any detective story!

Let us begin with a few details of the experiment itself. The basic idea was to use ^{37}Cl to detect the neutrinos. The reaction that was used is:

$$\nu + {}^{37}\text{Cl} \Rightarrow e^- + {}^{37}\text{Ar},$$

which has a threshold energy of 0.8 MeV (meaning that this reaction will not take place if the energy of the neutrino is less than 0.8 MeV). The detector was a steel tank of the size of an Olympic swimming pool, containing about 400,000 litres of C_2Cl_4 (a cleaning liquid used by your dry cleaner) more than a kilometre below the ground in the Homestake Gold Mine in South Dakota, USA (Fig. 5.14). The tank had to be buried deep under ground to minimize background events from cosmic rays. The aim of the experiment was to detect and count the number of Argon atoms produced in the tank. But this is easier said than done. Bahcall and his colleagues calculated the number of neutrinos of different energy that the Sun produces using a detailed computer model of the Sun. They also calculated the number of radioactive argon atoms (^{37}Ar) that these solar neutrinos would produce in the tank containing chlorine-based cleaning fluid. These calculations suggested just a few atoms of ^{37}Ar would be produced in this huge tank containing 100,000 gallons of chlorine. Although many ridiculed the idea of detecting just a few atoms, Davis and his experimental colleagues were very confident. Every few months, Davis and his collaborators extracted a small

Fig. 5.14 The chlorine experiment one kilometre below the ground in the Homestake Gold Mine in South Dakota, USA. The steel tank contains 400,000 litres of C_2Cl_4. When a neutrino interacts with a chlorine nucleus, a radioactive nucleus of argon is produced. The name of the game was to extract the few argon atoms out of a total of 1030 chlorine atoms in the tank!

sample of radioactive argon (^{37}Ar) from the tank; typically the number of argon atoms was of the order of 15. *Yes, 15 argon atoms out of a total of more than* 10^{30} *atoms in the tank*! Controlled experiments were performed to show that the radioactive argon produced by the neutrinos is extracted with more than 90 percent efficiency. As John Bahcall remarked, Davis 'had to be spectacularly clever' to be able to do this. And he was! The argon produced in the tank was separated chemically, purified and counted in low background counters. The results are expressed in terms of the Solar Neutrino Units, *SNU, which is the product of a characteristic calculated solar-neutrino flux* (cm^{-2} s^{-1}) *times a theoretical cross-section for neutrino absorption* (cm^2). An SNU has, therefore, the units of events per target atom per second and is chosen for convenience equal to 10^{-36} s^{-1}.

The first results were announced by Davis in 1968. *The chlorine detector had detected the solar neutrinos*! But the number of radioactive argon atoms produced by the neutrinos was only about one-third as many as were predicted. While there was considerable jubilation about the detection, some were irritated and concerned at the lack of agreement between the observations and the predictions.

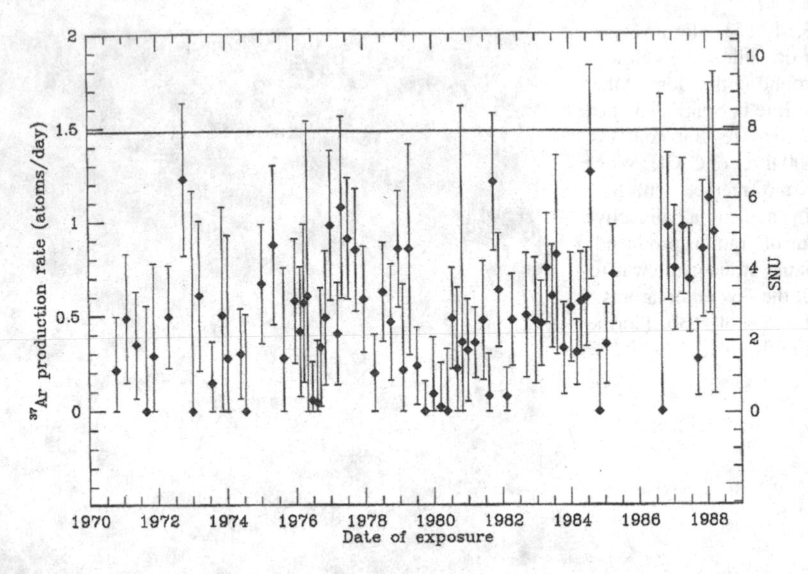

Fig. 5.15 The experimental data from the chlorine detector obtained over a period of nearly twenty years. From the average rate at which ^{37}Ar is produced, one calculates the average capture rate of the solar neutrinos. This capture rate is (2.05 ± 0.03) SNU. The line at 7.9 SNU across the top of the figure represents the prediction of the Standard Model of the Sun. This figure has been reproduced from *Neutrino Astrophysics* by John Bahcall, published by Cambridge University Press (1989)

What was Wrong?

Well, there were three possibilities, and all of them were suggested.

1. Perhaps the theoretical calculations were wrong. This might either be due to the predicted number of neutrinos being wrong or the calculated production rate of argon atoms in the detector being wrong.
2. Perhaps the experiment was wrong.
3. May be something is wrong with fundamental physics?

These reactions were surprising. As Bahcall himself said, given the complexity of the experiment, as well as the various uncertainties in the theoretical predictions, one should have been thrilled with the agreement within a factor of three. As it turned out, great progress was eventually made because neither the theorists nor the experimentalists were willing to concede that they had made a mistake.

Over the next two decades, Bahcall and his colleagues all over the world refined the theoretical calculations. The data used in these calculations were improved so that predictions became more precise. The computer model of the Sun was checked again and again, and no error was found. Similarly, the probability of detection of the neutrinos by Davis' tank was repeatedly checked, and no error was found.

On the experimental front, Davis and his colleagues continued to improve the sensitivity of the experiment. Great effort was made to understand the errors in the experiment better. But the discrepancy between the theoretical prediction and the experimental result did not go away. Figure 5.15 shows the observed rate over two decades. While the discrepancy appeared to be statistically significant, one might feel that by astronomical standards the *agreement* is quite satisfactory. The question is how seriously should one take the theoretical predictions? The production rate of neutrinos depends on a large power of the central temperature. An agreement within a factor of three tells us that we have, in a sense, measured the central temperature of the Sun to within a few percent. Eddington would have been thrilled with this. So why were the astronomers getting worked up over the discrepancy? Physicists continue to have an old-fashioned (and slightly contemptuous) attitude about the accuracy of astronomical results. They tend to feel that astronomical results are only accurate to within an order of magnitude (that is, a factor of ten). But this stereotype view can often be wrong. In this case, for example, astronomers felt that they knew the conditions inside the Sun to better accuracy than *99.9 percent*. Extraordinarily, the astronomer's *standard model* of the Sun is consistent with observations to an accuracy of better than 99.9 percent, but one had to wait for thirty years to establish this. But way back in 1968 both Davis and Bahcall were convinced that the discrepancy between the experimental result and the theoretical prediction was real, and that it should be accounted for. As we shall see in Chap. 7, this insistence led to several new experiments.

But we are jumping the story! The first results were announced by Davis in 1968. And it took twenty years to be really sure about the results of the experiment, as well as the theoretical modelling of the Sun. If one rules out a significant error in the theoretical modelling of the Sun, as well as the experiment, then we are left with only the third alternative mentioned above. As Sherlock Holmes would have said, 'My dear Watson, if all other alternatives have been ruled out, then the remaining one, however implausible, must be true'. The third alternative is that fundamental physics of elementary particles needs a revision.

Something Happened on the Way to the Earth!

One person did not wait for twenty years to come to this conclusion. Soon after the discrepancy surfaced, Bruno Pontecorvo (an Italian physicist working in Moscow, see Fig. 5.16.) and his colleague Vladimir Gribov proposed in 1969 that the results obtained by Davis pointed to the need for a drastic revision of elementary particle physics. Specifically, they made the radical suggestion that the discrepancy is due to the fact that *neutrinos suffered from personality disorder* (as Bahcall put it); they *oscillated* between three possible *incarnations—electron neutrino, muon neutrino and tau neutrino*. As the original electron neutrinos journey to the Earth, they change their identity back and forth. Finally, when they arrive at Earth only a fraction of them would be wearing the same hat as when they started. Since the experiment by Davis recognized only electron neutrinos, the observed discrepancy could be reconciled. Imagine that Dr Jekyll clones himself, and a hundred of these clones set out from the

Fig. 5.16 Bruno Pontecorvo

Sun. While travelling to the Earth for a party, they change their identity periodically between that of Dr Jekyll and Mr Hyde. So when they arrive at the party, there are likely to be fifty Doctor Jekylls and fifty Mister Hydes. Unfortunately, only Dr Jekyll will be allowed into the party. So their number at the party will be half of the number that set out from the Sun.

The suggestion made by Pontecorvo was as simple as that! As we shall discuss in Chap. 7, Pontecorvo had suggested way back in 1957 that neutrinos may oscillate between the three *flavours*, as they are referred to. But if this was the correct explanation for the discrepancy observed by Davis then it calls for a major revision of fundamental physics. Thirty-two years later, on 18 June 2001, a group of experimenters made a dramatic announcement: they had solved the solar neutrino mystery. Pontecorvo was right! Neutrinos do oscillate in flavour. Fundamental physics does need a revision!

But before I tell you that story, I must first tell you how astronomers convinced themselves that their computer model of the Sun was remarkably good; so good that its predictions agreed with the conditions in the interior to better than 99 percent accuracy. That is another interesting story and that will be the theme of the Chap. 6.

Chapter 6
Sounds of the Sun

At first sight it would seem that the deep interior of the Sun and stars is less accessible to scientific investigation than any other region of the universe. Our telescopes may probe farther and farther into the depths of space; but how can we ever obtain certain knowledge of that which is being hidden behind substantial barriers? What appliance can pierce through the outer layers of a star and test the conditions within?

Sir Arthur Eddington
The Internal Constitution of the Stars

This chapter is devoted to the discussion of some recent developments that allow us to 'pierce through the outer layers of a star and test the conditions within.' However, before we embark on this let us briefly recall what is known as the *Standard Model of the Sun.*

The Standard Model of the Sun

In Chap. 5, we discussed the pioneering experiment by Davis and his colleagues to detect the neutrinos from the Sun. We saw that they succeeded in this endeavour against all odds; however, rather than jumping with joy at having finally detected the elusive neutrinos from the Sun, Davis and Bahcall were bothered by the *discrepancy* between the predicted and the observed flux of solar neutrinos. In 1968, when the first result from the chlorine detector was announced, many felt that Bahcall was reading too much into this apparent discrepancy. The relevant question was this: Is the Standard Model of the Sun accurate enough for this apparent discrepancy to be taken seriously? After all, the predicted flux of solar neutrinos depends on the details of the Standard Model in a sensitive manner. Initially, only Bahcall and a few others thought that the Standard Model was accurate enough to warrant taking the discrepancy seriously. But by 1990, there was direct experimental evidence to show

G. Srinivasan, *What are the Stars?* Undergraduate Lecture Notes in Physics,
DOI: 10.1007/978-3-642-45302-1_6, © Springer-Verlag Berlin Heidelberg 2014

that the Standard Model of the Sun was, indeed, accurate to about 0.1 percent. This suggested that one should take the predicted neutrino flux seriously. If one could also be equally confident about the theoretical predictions regarding the interaction of the neutrinos with the detector, and if one could be sure about the errors in the experiment, then one had to take seriously the discrepancy between the results obtained by Davis and Bahcall. This chapter is devoted to the recent developments that enabled one to conclude that the Standard Model of the Sun was remarkably accurate. But before we tell that story, let us digress briefly to discuss what the Standard Model is all about.

We shall not pause to dwell on the intricacies of the Standard Model, nor on the details of the calculational method, but it is good to have an idea of the kind of results churned out by this model. The final objective of the model is to predict, with as much precision as possible, the radial profiles of various quantities such as (i) energy production, (ii) temperature, (iii) density, (iv) electron density, etc. This is easier said than done! Scores of astrophysicists all over the world have been engaged in this effort for several decades. The calculation of the Standard Model involves a number of *approximations* concerning, for instance, the following:

1. Hydrostatic equilibrium
2. Energy transport
3. Energy generation by nuclear reactions
4. Abundance changes *caused* by nuclear reactions

A detailed model of the internal structure of the Sun requires many input parameters. The key ones are given below.

1. The *chemical abundance* of the Sun
2. The *opacity* of the solar material to the outward flow of radiation
3. The *equation of state*: the relation between the pressure, density and temperature

The Standard Solar Model is calculated using the best physics and input parameters available. During the last twenty-five years or so, many hundreds of improvements in the input parameters have been incorporated.

A landmark paper in this field is the monumental article by Bahcall and Ulrich, published in *Reviews of Modern Physics* (1988). Figure 6.1a–d, given below, are reproduced from this article. Notice that the energy-production peaks at a radius of 0.09 R, where the symbol, R, stands for the Sun's radius. Notice also that the values of the central temperature and density are, respectively, 15.6×10^6 K and 148 g cm^{-3}, remarkably close to what Eddington had estimated way back!

Recall that one of the primary objectives of such a model is to predict the flux of solar neutrinos. Accordingly, the model provides *tables* giving a detailed numerical description of the solar interior. Starting from the centre, it gives, as a function of the radius, *temperature, density, luminosity generated, chemical abundances and the flux of neutrinos produced in the various reactions* (mentioned in Chap. 5). Therefore, the prediction of the total neutrino flux is quantitatively reliable only to the extent the standard model accurately describes the density and temperature profile.

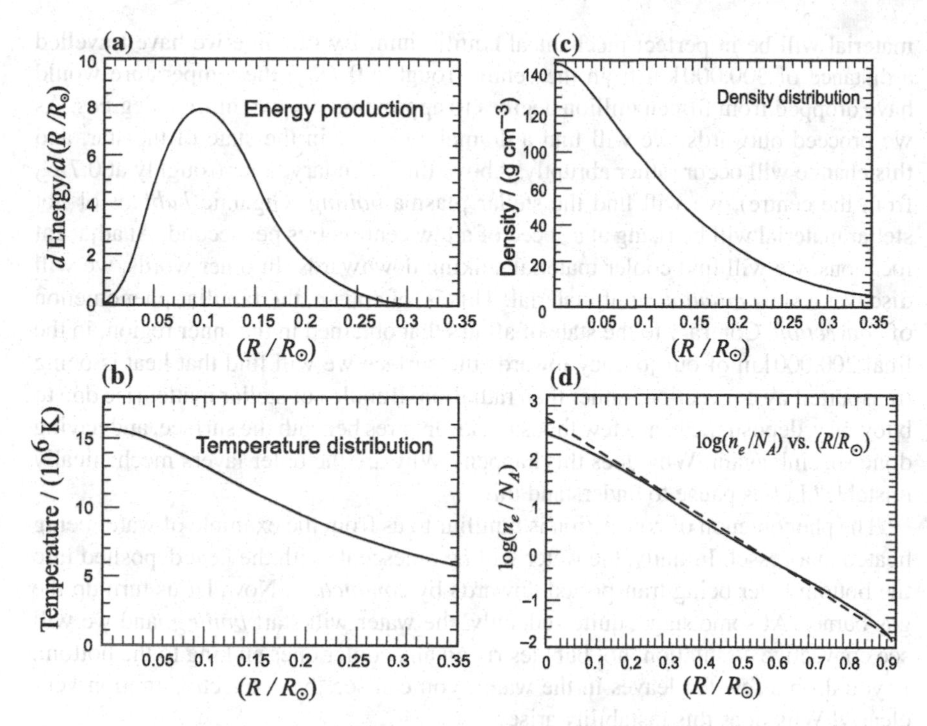

Fig. 6.1 Radial profiles of physical parameters. Figure **a** shows the fraction of the energy that is produced at each position in the standard solar model. Figure **b** shows the temperature profile. Figure **c** shows the density distribution. The *solid line* in **d** is the distribution of the electron density, n_e, divided by Avogadro's number, N_A, as a function of the solar radius. The *dotted line* is a theoretical fit to the electron density distribution. (This figure has been reprinted with permission from Bahcall, J.N. and Ulrich, R.K., *Reviews of Modern Physics*, volume **60**, 297 (1988). Copyright (1988) by the American Physical Society)

The Phenomenon of Convection

Before proceeding, let us first discuss a major departure of the structure of the Sun from what was envisaged by Eddington. In the discussion we have had so far, we have assumed that our Sun is in perfect hydrostatic equilibrium. Remember that Eddington had discarded the earlier suggestion of Lane that heat energy is transported outward by convection. Instead, he introduced the notion that heat is transported by radiation itself, and that the star would be in hydrostatic equilibrium. But what is the real situation in the Sun? Let us embark on a journey from the centre of the Sun towards its surface to get an overall picture of its internal structure.

As we proceed outwards from the core, we will find that the stellar material is chemically homogeneous and the temperature gradually decreases. We shall also find that the star is in *radiative equilibrium,* as desired by Eddington. In other words, the temperature gradient will be such that all the heat generated near the centre is transported outwards by radiation and there is no *piling up* of heat energy. The stellar material will be in perfect mechanical equilibrium. By the time we have travelled a distance of 500,000 km from the centre (roughly $0.7R_\odot$) the temperature would have dropped from fifteen million degrees to approximately two million degrees. As we proceed outwards, we will find a dramatic change in the state of the star, and this change will occur rather abruptly. Above this boundary layer (roughly at $0.7R_\odot$ from the centre), we will find the stellar plasma *boiling*. Gigantic *bubbles* of hot stellar material will be rising at a speed of a few centimetres per second. At adjacent locations, we will find cooler material sinking downwards. In other words, we will discover a slow *circulation* of material. This is, of course, the familiar phenomenon of *convection*. Contrary to the state of affairs that obtained in the inner region, in the final 200,000 km of our journey towards the surface we will find that heat is being transported by convection, rather than radiation. Bundles of stellar matter rise due to buoyancy, deposit the heat a few thousand kilometres beneath the surface, and having done so, sink again. Why does this happen? Why are the outer layers mechanically unstable? Let us pause to understand this.

The phenomenon of convection is familiar to us from the example of water being heated in a vessel. Initially, the water will be quiescent, with the heat deposited into the bottom layer being transported upwards by *conduction*. Now, let us turn up the gas burner. At some stage, quite suddenly, the water will start *boiling*, and we will see convective circulation, hot bubbles rising and cooler ones sinking to the bottom. If you drop a few tea leaves in the water, you can see the convective motion very clearly! Why does this instability arise?

Instead of discussing a liquid being heated (where heat is transported by *conduction*, under normal circumstances), let us go back to the discussion of stellar material where heat is transported by *radiation*. (Heat transport by *conduction* is not very effective in a stellar plasma.) Let us simplify it to a one-dimensional problem depicted in Fig. 6.2. The force of gravity acts downwards. Consequently, the pressure, density and temperature of the gas decrease as we go outwards in the radial direction. Imagine that an element of the plasma at a radial distance, r, is heated to a higher temperature than the surroundings due to some fluctuation. We want to know how this element will respond. A higher temperature implies the prevalence of higher pressure inside the element as compared to the surrounding; this follows from *Boyle's law*. This excess pressure inside the element will cause the plasma bubble to expand *until the pressure inside the element is, once again, equal to the pressure outside*. Thus, *pressure equilibrium* is established with respect to the surroundings. But since the plasma bubble we are discussing is *hotter* than the surroundings, it will have to be at a lower density than the surrounding for pressure equilibrium to be achieved. (Recall, in the case of an ideal gas, the pressure is $\propto \rho T$. Therefore, *for a given pressure, higher the temperature lower will be the density.*)

$$P_{\text{element}} = P_{\text{surrounding}} \Rightarrow ((\rho - \Delta\rho)(T + \Delta T)) = \rho T .$$

Consequently, the element we are discussing will rise due to *buoyancy force* to a new height, $r + \Delta r$. The question is whether it will continue to rise or sink back to the original depth after rising for a while. If it continues to rise, we have a convective

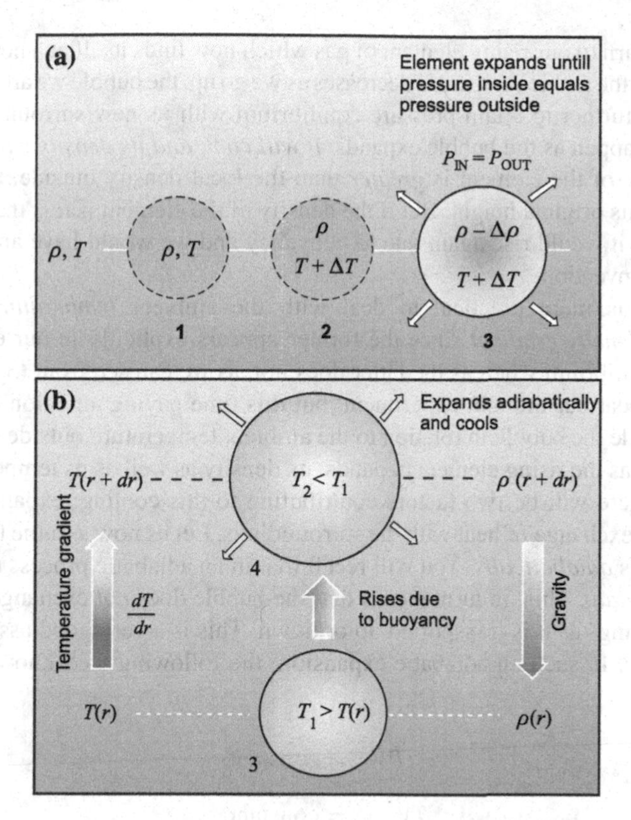

Fig. 6.2 Condition for convection to set in. **a** Consider a sample element of stellar plasma, labelled 1, at a certain radial distance from the centre where the local ambient density and temperature are $\rho(r)$ and $T(r)$, respectively. Imagine that due to some fluctuation this element is heated to a slightly higher temperature $T(r) + \Delta T$ (labelled 2). Since this will result in the pressure inside the element to be greater than the surrounding pressure, it will expand till the pressure inside, once again, becomes equal to the ambient pressure. As a consequence of this expansion, the density inside the sample element will be less than the ambient density (labelled 3). **b** Since the density of our sample element (stage 3) is less than the ambient density, it will rise due to buoyancy. Since the ambient density, temperature and pressure are all decreasing as we go upwards, it will continue to expand as it rises. As our bubble rises and expands, it will cool. If there is no heat exchanged with the surrounding during the expansion, then the amount of cooling will be determined by the law of adiabatic expansion. If our bubble (stage 4) continues to be hotter than the local surrounding despite it cooling due to expansion, then it will continue to rise and we shall have convection

instability. The French physicist Henri Bénard investigated this important problem around 1900 in a series of very beautiful experiments. The great British physicist Lord Rayleigh derived the condition for *stability* against convection and provided a detailed theory of the phenomenon. The German physicist and astronomer Karl Schwarzschild studied this problem in the context of gaseous stars. We shall now briefly discuss the condition derived by him for stability against convection. And let us permit ourselves to look at the problem in rather simplified terms.

Let us return to our rising element of gas which now finds itself at a new location, $r+\Delta r$. Since the ambient pressure decreases as we go up, the bubble we are discussing will expand further to attain pressure equilibrium with its new surroundings. Two things will happen as the bubble expands: *it will cool, and its density will decrease.* If the density of the element is *greater* than the local density outside, then it will sink back to its original height. But if the density of the element is *less* than the local density, then it would rise again due to buoyancy and we would have an instability signalling convection.

Often, it is more practical to deal with the ambient *temperature gradient* rather than *density gradient* since the former appears explicitly in our equation of radiative equilibrium whereas the latter does not, as we can see from (3.12). Let us therefore repeat our thought experiment, but this time paying attention to the temperature inside the bubble in relation to the ambient temperature outside the bubble. We said that as the rising element expands, its density as well as its temperature will decrease. There will be two factors contributing to this cooling: expansion of the element and exchange of heat with the surroundings. Let us now assume that the element expands *adiabatically.* You will recall that in an adiabatic process the *entropy remains constant.* This, in turn, means that the bubble does *not* exchange heat with the surroundings as it is transported up or down. This is a very good assumption in gaseous stars. In such an adiabatic expansion, the following well-known relations hold:

$$PV^{\gamma} = \text{constant}. \tag{6.1}$$

$$TV^{\gamma-1} = \text{constant}. \tag{6.2}$$

Here T is the temperature; P, the pressure; V, the volume, and $\gamma = c_P/c_V$, the ratio of the specific heat at constant pressure, c_P, to the specific heat at constant volume, c_V. Equation (6.1) tells us how the pressure will change in an adiabatic expansion, while Eq. (6.2) tells us the amount by which the gas will cool if it expands adiabatically.

Let us return to our plasma bubble which is expanding adiabatically as it rises due to buoyancy and cools in the process. If our element finds itself again *hotter* than the surrounding then it will continue to rise and we have an instability. Therefore, whether our element would continue to rise or not depends on the comparison between two *temperature gradients*. One is, of course, the actual temperature gradient that obtains in the star; the rate at which the temperature of the different mass shells decreases as one moves outwards from the centre. For a star in radiative equilibrium, we derived this temperature gradient in Chap. 3, Eq. (3.12). According to this equation given the rate at which energy flows outward through an imaginary shell of the star (the left-hand side of this equation), the *temperature gradient* or the *rate at which the temperature decreases as a function of the radial distance,* is determined by the opacity of the stellar material. This equation must be satisfied at every radius. Therefore, given the rate of energy production at the centre and the opacity of the stellar material, the temperature gradient adjusts itself to the required value everywhere in the star. This is the principle of radiative equilibrium. The other *gradient*, known as the *adiabatic temperature gradient*, is not really a spatial gradient of temperature. Rather, it is the rate at which the temperature of an element changes as it is transported adiabatically in a radial direction. Thus, the condition for convection to set in may be stated as follows:

Radiative temperature gradient > Adiabatic temperature gradient

Remember that if the *gradient* is more, then the temperature decreases more rapidly as we go outwards. The expression for the so-called adiabatic temperature gradient will be found in any standard book on fluid dynamics. We shall not attempt to derive it here. Instead, we shall merely state the final result, which was first obtained by Karl Schwarzschild. Stated in terms of the temperature gradient, convection can occur if the temperature falls with increasing height, and if the magnitude of the temperature gradient exceeds a critical value:

$$\boxed{\left|\frac{dT}{dr}\right|_{\text{rad}} > \frac{g}{c_p}.} \tag{6.3}$$

In Eq. (6.3), g stands for the acceleration due to gravity and c_p stands for the specific heat at constant pressure. Equation (6.3) is the famous criterion first derived by Karl Schwarzschild. The meaning of this inequality is as follows. The left-hand side is the rate at which the ambient temperature decreases as a function of the radial distance. The right-hand side is the rate at which the temperature of our sample element decreases as it moves up and expands adiabatically. The condition that our sample element is hotter than the ambient matter is given by the above inequality.

Why does Convection Occur in the Sun?

Let us now continue with the description of our journey from the centre of the Sun. We had reached a distance of 500,000 km from the centre. There we found a boundary layer above which convection was taking place. We conclude from the above discussion that this must be so because above this layer the radiative temperature gradient is steeper than the adiabatic value, which is the condition stated in Eq. (6.3). The question is why did the temperature gradient suddenly steepen and, at some level, become larger than the adiabatic value? After all, up to that point, the star was in hydrostatic equilibrium. The answer is to be found in (3.12) for radiative equilibrium we derived in Chap. 3, and which is reproduced below for convenience:

$$L(r) = -\frac{ac}{3\kappa\rho}\frac{dT^4}{dr}4\pi r^2. \tag{6.4}$$

Here $L(r)$ is the luminosity crossing an imaginary sphere of radius r. In other words, it is the amount of outward flowing energy crossing this imaginary surface per unit time. Let us recall the meaning of this equation. The basic assumption is that all the energy generated near the centre (per unit time) is transported outwards and eventually radiated away from the surface. There can be no piling up of the generated energy. Under the condition of radiative equilibrium, the net outward flux of radiation is determined by the temperature gradient and the *opacity* κ of the stellar material. Therefore, given the energy generation rate at the centre, the above equation says that if the opacity of stellar material increases, then the temperature gradient would also have to increase proportionately (because the opacity is in the denominator and the temperature gradient is in the numerator). Therefore, the onset of convection must be related to an increase in the opacity of the outer layers of the star, and it is to a discussion of this we now turn to.

Increased Opacity in the Outer Layers

At the very high temperature that obtains in the inner regions, hydrogen is completely ionized and the heavier elements are stripped of all but their innermost electrons. As we discussed in Chap. 3, the main mechanisms contributing to opacity are photo-electric absorption by the innermost electrons still bound to the nuclei of the heavy elements, and electron scattering. We had remarked that *bound–bound transitions* are not a significant source of opacity in the inner regions of the star. Recall that these are transitions in which an electron bound to a heavy ion absorbs a photon and jumps from an occupied level to an unoccupied level (see Fig. 3.7). However, we had pointed out that these transitions could become important when the temperature drops to approximately one million degrees. As we journey from the centre to the surface, we find ourselves in a region where the temperature has dropped to a mere two million kelvin. At this temperature, one will encounter ions of C, N, O, Ca, Fe and so on, with a few more bound electrons than they had near the centre (species of ions that we did not find near the centre). These ions will provide new channels for absorption of the ambient soft x-ray and far-ultraviolet photons which, in turn, will significantly increase the opacity. Such an increase in opacity will result in an increase in the temperature gradient.

The Negative Ion of Hydrogen

As we go out even further, one ion plays a central role in the dramatic increase in the opacity of the outer layers, particularly in the cooler stars like the Sun, and that is the *negative ion of hydrogen*—H^- (a hydrogen atom with *two* electrons!) The *ions* you would be familiar with are positively charged; the number of electrons orbiting the nucleus is *less* than the number of protons in the nucleus. Here we are talking about a negative ion, with the number of electrons being *more than* the number of protons. It is likely that you may not have heard of this before. Let us, therefore, digress a little since the story of is fascinating.

Soon after the advent of the quantum theory of atoms, it was pointed out by some perceptive physicists that a hydrogen atom with two electrons could be a bound state. But the proof came only in 1929, when Hans Bethe published a seminal paper (the same Bethe who worked out the details of the energy production in the Sun). is a weakly bound iȯn with a binding energy of 0.75 eV (recall that the binding energy of the neutral hydrogen atom is 13.6 eV). What this means is that a photon with energy of 0.75 eV can knock out the second electron. And that is precisely why it is of such importance in astrophysics. Around 1940, an astronomer by name Wildt realized that ions could contribute significantly to the opacity of stellar atmospheres. For one thing, neutral hydrogen is abundant in the outer layers. And there are plenty of free electrons. Thus it is the ideal environment for the formation of bound ions—a neutral hydrogen atom capturing a second electron. At the same time, there are plenty of photons of the right energy to destroy these ions by detaching the second electron. As we approach the *surface* (or, more correctly, the *photosphere*), the temperature would have dropped to around ten thousand degrees. Accordingly, the energy of most of the photons would also have decreased. Remember that at a temperature of ten thousand degrees, the *peak* of the black body spectrum would shift to photon energy of the order of one electron volt. Most neutral atoms and positive ions are *transparent* to photons of such energy and, therefore, do not contribute to the opacity. This is because their first photoelectric absorption occurs at 4 or 5 eV, if not larger, energy; our tired photons do not have enough energy to kick an electron to a higher energy level. This is where the H^- ion comes into play. In the infrared and visible wavelengths (which are roughly in the energy range 0.75–4 eV), the H^- ion is *not* transparent. Even our *tired photons* will have enough energy to knock off an electron from the H^- ions (recall that the binding energy is only 0.75 eV). Thus, the negative ion of hydrogen is a dominant contributor to the opacity in the outer layers of stars like the Sun. Subrahmanyan Chandrasekhar was quick to appreciate this, and went on to contribute significantly to the story of the H^- ion, both in terms of the quantum mechanics of this ion and its implication for the stellar atmospheres of cool stars. But we have digressed too far. Let us return to our discussion of convection in the outer regions of the Sun.

Until we reached a distance $\sim 0.7 R_\odot$, we saw that the star was in hydrostatic and radiative equilibrium. Beyond this point, the opacity of stellar matter increased significantly. This, in turn, resulted in an increase of the temperature gradient. At some stage, the temperature gradient will exceed the adiabatic value, and convection sets in. Radiation is no longer able to transport the heat in a diffusive manner, as it had done all the way from the centre. Instead, from here on heat is transported by convection. Bubbles of solar plasma carry the heat as they buoyantly move up and deposit it just below the surface. Having done so, the cooler bubbles sink back to the lower boundary of the convection layer. Figure 6.3 shows the internal structure of the Sun, taking into account convection in the outer regions. Some years ago, it

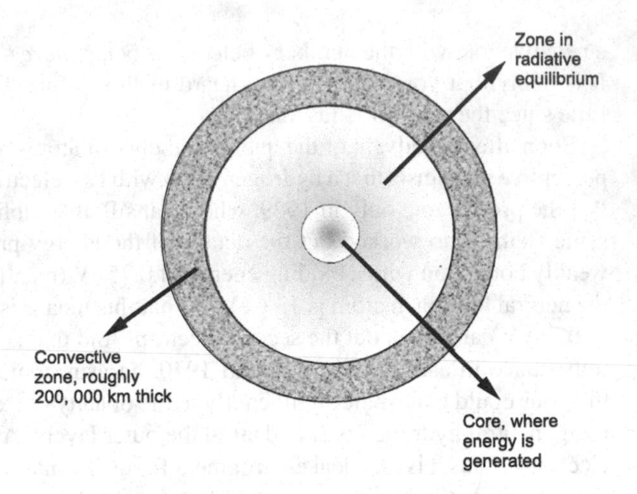

Fig. 6.3 The internal structure of the Sun. Much of the interior of the Sun is in hydrostatic equilibrium. In this region, the energy generated in the core is transported outwards by radiation itself. But convection prevails in the outer regions. What causes this is the increased opacity of the outer layers. Increased opacity causes a steepening of the ambient temperature gradient. When the temperature gradient exceeds a critical value, convection sets in

was believed that the convective layer is only about 60,000 km thick. But according to current thinking, it is much thicker, and extends to a depth of nearly 200,000 km.

Evidence of Convection: Granules and Supergranules

So far we have discussed convection as an orderly phenomenon with *blobs* of hot stellar plasma rising due to buoyancy, depositing the excess heat in the outer layers and sinking due to gravity. These blobs, in a sense, played the role of the *molecules* in a microscopic picture. But this is a highly oversimplified picture. The highly *ordered* picture of convection is *messed up* by chaotic motion or *turbulence*. The physics of fully developed turbulence is still in its infancy. The mathematical description of turbulence is very complicated and not amenable to convenient approximations. Although experimental investigations of turbulence are now very mature and sophisticated, it is rather hard to simulate the conditions that obtain in astrophysical contexts; one has to resort to computer simulations to gain insights. A fair amount of progress has been made in this direction, but we shall not pause now to summarize that. For our purpose, it will suffice to describe the effect of turbulence on convective motions in the following terms.

At the *base* of the convection zone, we would be right in describing the convective motions in terms of fat columns of hot plasma rising, interspersed with columns of cooler plasma sinking. But this picture breaks down as we approach the top of the zone. Here, the rising columns break up into giant *eddies,* roughly 30,000 km wide. The hot plasma rises near the centres of these eddies, flows horizontally to the periphery and sinks there. Imagine a swimmer diving from a high board. He will pierce the water, go down for a while and then start to rise. Just before he surfaces, you will see a swell of water just over his head. This swell will disperse horizontally.

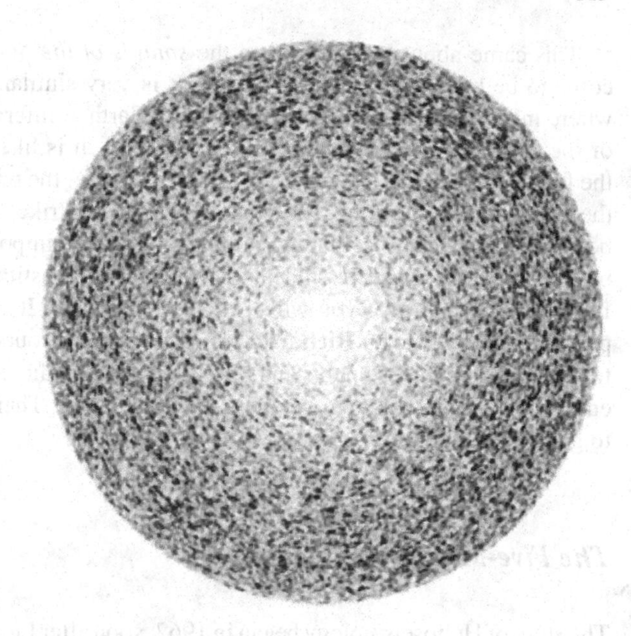

Fig. 6.4 Granulation of the surface of the Sun. The dark features seen in this resolution are known as *supergranules* (Courtesy of *SOHO MDI/SOI*). They are roughly 30,000 km in size and last for about a day. So the pattern will be changing on the timescale of a day

This is exactly what we find happens in the giant eddies, known as *supergranules,* that are seen on the surface of the Sun. Figure 6.4 shows the surface of the Sun peppered with these supergranules. These eddies are not permanent. Typically, they last for about a day. Thus the pattern of convective cells keeps on changing.

The story does not end there. If we look closely we will find that these giant cells consist of smaller eddies, which, in turn, consist of even smaller eddies. Figure 6.5 shows the pattern of cells with a characteristic size of about 1,000 km. These are known as *granules*, and last for only a few minutes. We shall not pause to discuss the many interesting phenomena associated with granules and supergranules. Our intention was merely to draw attention to these and point out that these are manifestations of the large-scale convective motions in the outer layers.

Sounds of the Sun

Let us now return to the main objective of this chapter. We began by asking whether the Standard Model of the Sun was accurate enough for us to take seriously the *discrepancy* between the observed and predicted flux of solar neutrinos. We mentioned that although initially only Bahcall and a few others thought so, by 1990, there was direct experimental evidence to show that that the Standard Model was accurate to about 0.1 percent. Astronomers had been able to *pierce through the outer layers of a star and test the conditions within.* Eddington would have been thrilled.

This came about by listening to the *sounds of the Sun*. And the technique has come to be known as *Helioseismology*. It is very similar to terrestrial seismology where information about the interior of the Earth is inferred by a systematic study of the slight motions of the surface of the Earth. It is like striking a bell and using the frequencies of the sound waves produced to infer the properties of the material of the bell. Actually speaking, one does not have to its strike the Sun to make it oscillate because its surface is oscillating spontaneously. This important discovery was made by Robert Leighton and his students at the California Institute of Technology in 1962. I am sure that many of you will recognize his name. It was Robert Leighton who persuaded the legendary **Richard Feynman** to give a course of **Lectures on Physics** to the undergraduate students and published them in three volumes. If you have not encountered these volumes, get them, and read them! There is simply no better way to get excited about how Nature works.

The Five-Minute Oscillations

The story of Helioseismology began in 1962, soon after Leighton and his students had discovered the *supergranules* we mentioned above. Leighton had built a sophisticated instrument to measure small velocities in these supergranules. The idea was to pick one of the absorption lines in the spectrum of light from a particular spot in the Sun and determine its wavelength very accurately. Depending on whether the absorbing material is moving towards us or away from us, the wavelength of the absorption feature will be shifted to a shorter wavelength (*blue shift*) or longer wavelength (*red shift*), respectively, in comparison with what one would expect in the absence of any motion. This is just the familiar Doppler effect. You will remember that the fractional shift in the wavelength is given by

$$\frac{\Delta\lambda}{\lambda} = \frac{v}{c}, \tag{6.5}$$

where v is the component of the velocity of the source in the line of sight to the observer and c is the velocity of light. The *patch* of Sun that Leighton was looking at was large enough to include many *granules,* each behaving independent of the other—some rising and some sinking (see Fig. 6.5). Therefore, he should have found purely chaotic motion, with any systematic velocities at the granular scale cancelling out at a larger scale. Instead, he found that the large *patch* he was looking at was *oscillating* intermittently with a period of the order of 5 min and velocity of the order 0.5 km s^{-1}.

Let us pause to appreciate the accuracy of the measurements. Since the velocity of light is 300,000 km/s, a velocity of a mere 0.5 km s^{-1} will result in an incredibly small wavelength shift, as implied by the relation (6.5). It would be quite a challenge to detect this even if the spectral line being tracked was extremely narrow (in wavelength). Unfortunately, absorption features in astronomical sources tend to be very

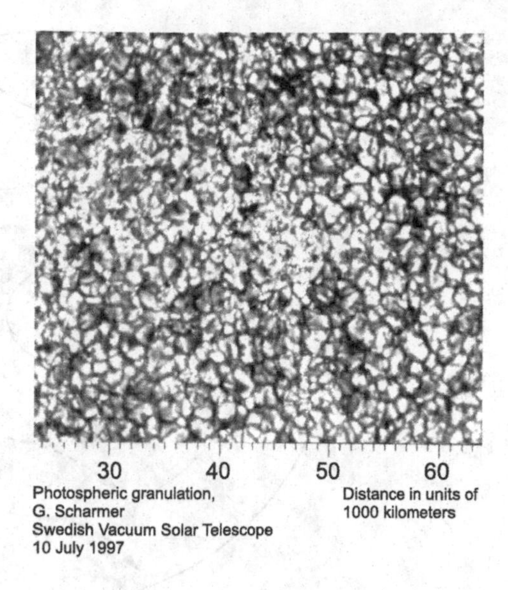

Fig. 6.5 If one looks carefully one will discover that supergranules themselves have a granular structure, as may be seen in this photograph. These granules have a characteristic size of the order of 1,000 km, and last only for a few minutes. (Courtesy of the Swedish Vacuum Telescope)

broad. Therefore, detecting small velocities, resulting in small shifts of broad lines, is always a Herculean task. But Leighton was well known for his clever experiments.

The Ringing Sun

Is the Sun's surface really oscillating? Why?

The correct explanation took quite some time in coming, and was advanced many years after Leighton's discovery, by **Roger Ulrich** (1970) and independently by **Robert Stein** and **John Leibacher** (1971). They argued that the surface oscillations of the Sun were caused by sound waves in the convection zone. According to them, the Sun acts as a resonant cavity, with sound waves known as *p-modes* (or pressure oscillations) trapped between the solar surface and the lower boundary of the convection zone. Sound waves represent the propagation of pressure oscillations, or in more familiar language, the propagation of periodic local compression and rarefaction of the gas. In sound waves, there is no net motion of the gas itself; sound waves represent the propagation of pressure oscillations. Ulrich and the duo, Stein and Leibacher, argued that sound waves in the Sun are trapped in the convective zone for the following reason. As these waves move outwards, they are reflected back near the solar surface due to the sharp density gradient. And as they move inwards, they are *bent back* or *refracted* due to the increasing speed of sound. The increase of sound speed is a direct consequence of the increase in the temperature of the stellar plasma as one goes deeper. Let us try to understand this. As mentioned above, sound waves are propagation of pressure oscillations. One normally assumes that the compression and rarefaction of the gas occurs *adiabatically*. This means that

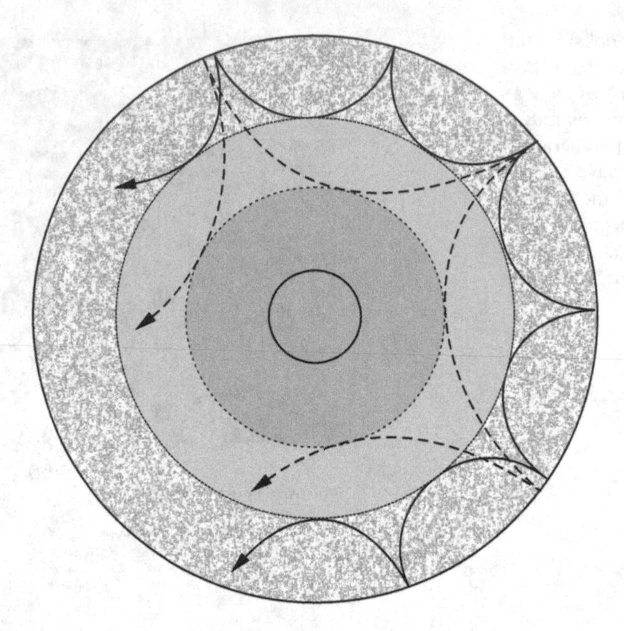

Fig. 6.6 A schematic representation of sound waves trapped inside the Sun. As these waves try to escape from the Sun, they are reflected inwards due to the sharp density gradient near the solar surface (*mirror effect*). Then, as they propagate inwards, they are bent back or *refracted*. This happens because the speed of sound inside the Sun increases as one goes deeper, a direct consequence of the increase in temperature as one goes deeper. The Sun thus acts as a resonant cavity of standing waves of various wavelengths. Notice that waves with longer horizontal wavelength penetrate deeper into the Sun

the entropy remains constant during the compression of a local element of gas, that is, no heat is exchanged with the surroundings. Under these conditions, the speed of sound is given by:

$$c_s = \left(\frac{\partial P}{\partial \rho}\right)_s^{\frac{1}{2}}, \tag{6.6}$$

where the subscript on the right-hand side signifies that the derivative is taken with the entropy s remaining constant. As we know, in an adiabatic process, the pressure and density are related by $P = K\rho^{\gamma}$, where K is a constant and $\gamma = \frac{5}{3}$ for an ideal gas. (This is just another way of writing Eq. (6.1). The volume term has been brought to the right-hand side and its reciprocal written in terms of the mass density ρ.) Differentiating, we get $\partial P/\partial \rho = 5/3\ (P/\rho)$. Recall, for an ideal gas $P = \rho kT/\mu m_H$. Therefore, $(\partial P/\partial \rho) \propto T$. It follows from that the speed of sound is $\propto \sqrt{T}$. Since the temperature increases as we go inwards, the sound speed also increases. This is the reason why sound waves are bent back as they try to propagate inwards; it is just the familiar phenomenon of *refraction*. The net result of the sound waves being reflected near the surface due to the mirror effect, and bent back in the convection zone due to refraction, is that they bounce back and forth in this spherical shell. This is depicted in Fig. 6.6.

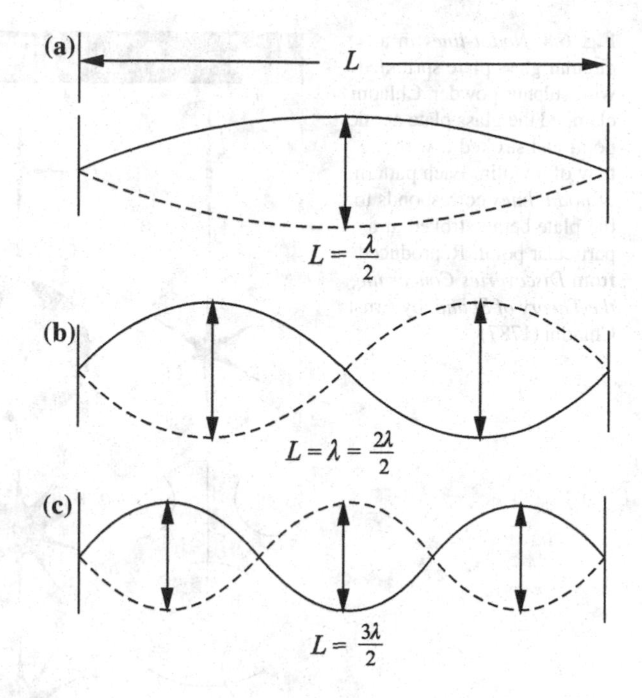

Fig. 6.7 Standing waves in a stretched string: locations where the string is stationary are called *nodes*

Clearly, for this phenomenon to be long-lived, the waves trapped in this region must form a standing wave pattern. Therefore, the idea was that the observed motions of the Sun's surface, such as discovered by Leighton, result from a superposition of several resonant modes with different periods and horizontal wavelengths. As we shall presently see, it is a superposition of not just a few modes but of *millions of resonant modes!*

Nodes, Nodal Lines and Nodal Surfaces

Before proceeding further, let us review some relevant aspects of *resonant* or *normal modes* of vibrations. Let us first discuss a one-dimensional problem, namely the vibration of a string.

Consider a string of length L-that is fixed at both ends, and excite it by plucking. It is clear that for a standing pattern to occur, the wavelength of the excitation, λ, must satisfy the condition $L = n\lambda/2$, where $n = 1, 2, 3 \ldots$, as shown in Fig. 6.7. The mode with $n = 1$, is known as the *fundamental mode*, while modes with $n = 2, 3, \ldots$ are known as *overtones* or *harmonics*. You will see Fig. 6.7 that in all the overtones there are points in the string where the amplitude of vibration is zero. These points are known as *nodes*. These are special points in the standing-wave pattern where the string is stationary. At all other points you will find the string

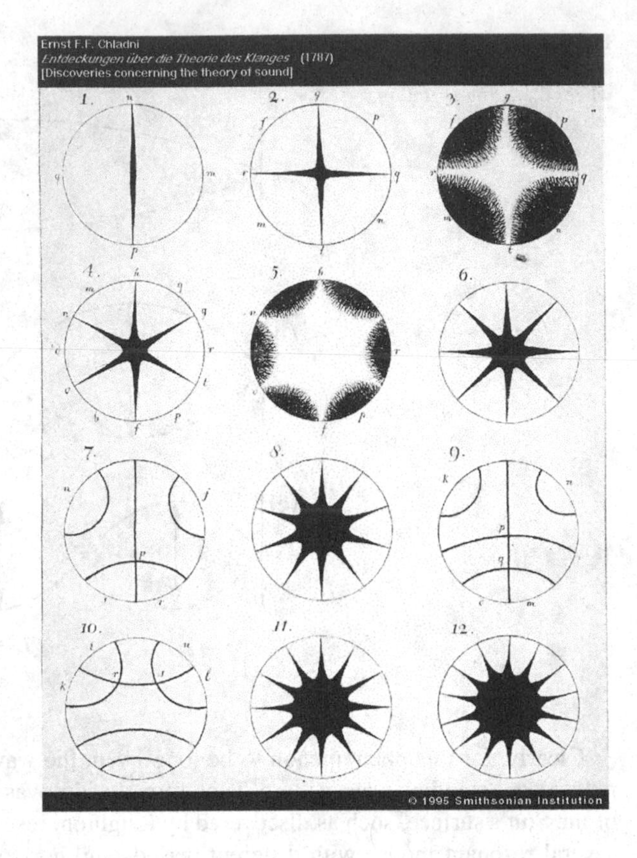

Fig. 6.8 *Nodal lines* in a circular glass plate sprinkled with sulphur powder. Chladni clamped the glass plate at one point and stroked it with the bow of a violin. Each pattern of *nodal lines* corresponds to the plate being stroked at a particular point. Reproduced from *Discoveries Concerning the Theory of Sound*, by Ernst Chladni (1787)

going up and down, with different amplitudes. The locations where the amplitude is a maximum are called *antinodes.* But at the *nodes* the string will be motionless.

Now let us go to vibrations in two dimensions, like waves in a pond or vibrations of the membrane of a drum. One of the most famous investigations of this phenomenon was carried out by **Ernst Chladni** in 1787. He clamped a glass plate in one corner, sprinkled sulphur powder on its surface and stroked one of the free edges at some particular point with the bow of a violin. What he saw was remarkable. The vibrations of the plate caused the sulphur particles to dance around, until they eventually settled at points on the plate that were stationary. These are the *nodal lines.* These are the two-dimensional analogues of nodes in a vibrating string. Everywhere else, the glass plate is moving up or down. But along the nodal lines, the vibrations due to criss-crossing waves that have been reflected repeatedly off the edges, cancel out. You will see in Fig. 6.8, the remarkable regular patterns formed by the nodal lines in a circular glass plate. The different patterns correspond to different tones of the vibrating plate. Such figures have now come to known as *Chladni figures.*

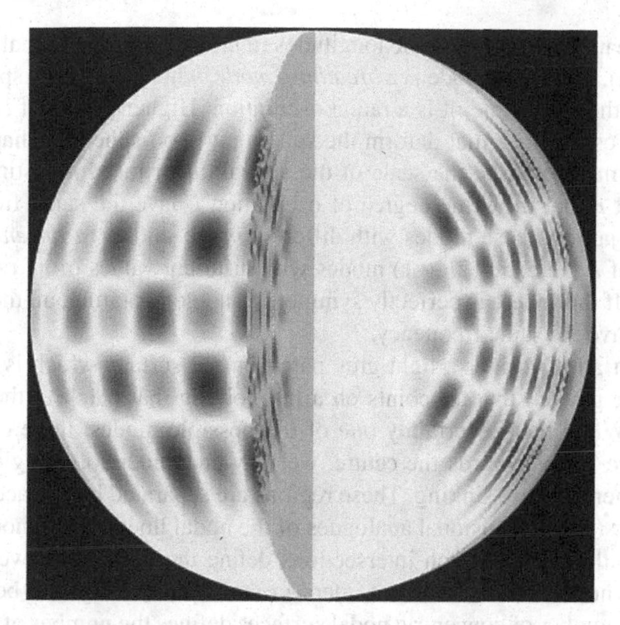

Fig. 6.9 *Nodal lines* and nodal surfaces in three dimensions. This figure shows a computer simulation of a three-dimensional gas sphere vibrating in one of its normal modes. Concentrate first on the patterns on its surface. The *light-grey* patches represent areas that are rising, while the very dark patches represent sinking areas. The sphere is stationary between these two areas; these are the nodal lines. Clearly, the nodal lines are latitudes and longitudes on the sphere. But the surface pattern does not reflect what is happening inside the sphere. Consider a *light grey* patch on the surface, representing a rising area. This does not mean that all points on a line joining this patch to the centre are also rising. This may be seen in the cut-out of the sphere. As one moves inwards, one will alternately encounter rising and sinking regions, separated by a thin layer which is stationary. Thus, there are nodal surfaces, in addition to *nodal lines*. As may be seen, these nodal surfaces are concentric spheres. In this particular normal mode, there are 16 nodes along the equator; 20 nodes along a longitude and 14 nodes in the radial direction! [Courtesy of Wikipedia, The Free Encyclopedia. Helioseismology: http://en.wikipedia.org/w/index.php?title=Helioseismology\&oldid=299610276

Vibrating Spheres

Now, let us go to three dimensions and consider the vibrations of a gas sphere. It is easy to anticipate that the spectrum of vibrations will be much richer in three dimensions. It is so rich that we can only study it with the help of computers.

Figure 6.9 shows a cut-out of a gas sphere pulsating in a particular mode. In this computer-generated picture, the light grey areas are rising and the dark grey areas are sinking. At the boundary between these two areas, the sphere is stationary. These are the nodal lines. As may be seen in Fig. 6.9, the nodal lines are the latitudes and longitudes on the sphere. Therefore, to characterize a mode, we have to specify the number of nodes, m, as we go around the equator (azimuthal angle φ in the usual spherical polar coordinate system), as well as the number of nodes, l, as we go along any of the great circles defining the longitudes (angle, θ, in the spherical polar coordinate system). The $l = 0$ mode is a *breathing mode*, where the whole sphere moves in and out at the same time; it is a radial oscillation. Higher values of l correspond to non-radial oscillations that deform the sphere into non-spherical shapes. Higher the degree, smaller will be the scale of the spatial distortion. Interestingly, modes with different m but the same degree of oscillation l have the same frequency; in the technical jargon, these modes with different m are said to be *degenerate*. For a given value of l, there are $(2l + 1)$ modes with different values of m, ranging from $-l \leq m \leq l$. If the sphere is perfectly symmetric and non-rotating, all these $(2l+1)$ modes will have the same frequency.

This is not all. Consider a light grey patch on the surface which is rising. Can we take it for granted that all points on a line joining that patch to the centre are also rising? While that is certainly *one* of the possible modes, there can be other modes. As we move towards the centre, we may encounter some regions that are rising and others that are sinking. These regions are separated by surfaces known as *nodal surfaces*(two-dimensional analogues of the nodal lines). Some nodal surfaces will intersect the surface. Such intersections define the nodal lines we mentioned above. Other nodal surfaces will not intersect the surface. They will be concentric spheres. The number of concentric nodal surfaces defines the number of nodes, n, in the *radial direction*. Therefore, to characterize a mode in three dimensions, we have to specify three numbers:

1. The *radial order* n, specifying the number of radial nodes.
2. The *degree*, l, the number of nodes along the meridian.
3. The *azimuthal order*, m, the number of nodes along the latitude.

This may sound very familiar to those of you who have studied some quantum mechanics. You will recall that three *quantum numbers* (n, l, m) are needed to define the wave function of the various levels of, for instance, a hydrogen atom. Not surprisingly, the three quantum numbers have the same meaning as in the present problem. They specify the number of nodes in the *wave function* in the radial, polar and azimuthal directions, respectively (Fig. 6.10).

The Frequency Spectrum

Let us first consider modes that differ in the number of nodes l along a longitude, but which have the *same number of radial nodes, n*. Modes whose degree of oscillation, l, is greater, will have higher frequencies. This is not difficult to understand. Recall that a larger value for l means a greater number of nodes. As the number of nodes increases, the wavelength of the excitations decreases and consequently the frequency increases. You will recall that this is the same with vibrating strings; the frequency of vibration of the string increases as the number of nodes increases. Therefore, a plot of frequency (on the Y-axis) versus degree of oscillation of modes (on the X-axis) will look like a *string of pearls,* sloping upwards as we go to a higher degree of oscillation (see Fig. 6.11). Remember, all this holds for a fixed n.

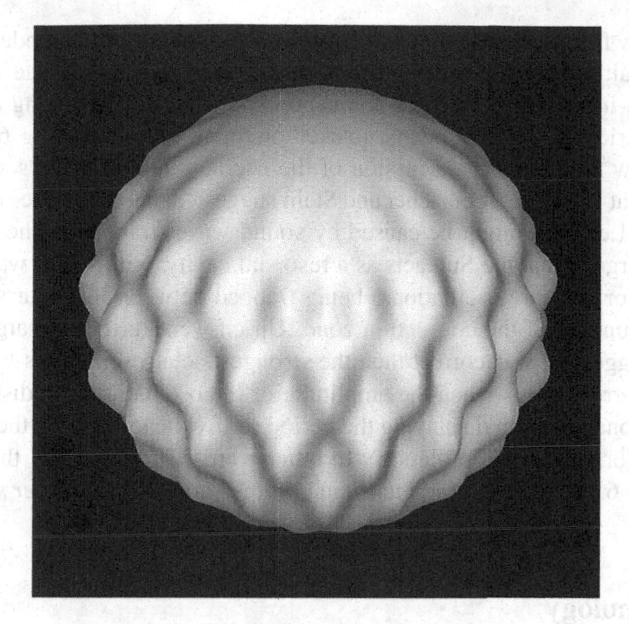

Fig. 6.10 Computer simulation of non-radial oscillation of a sphere

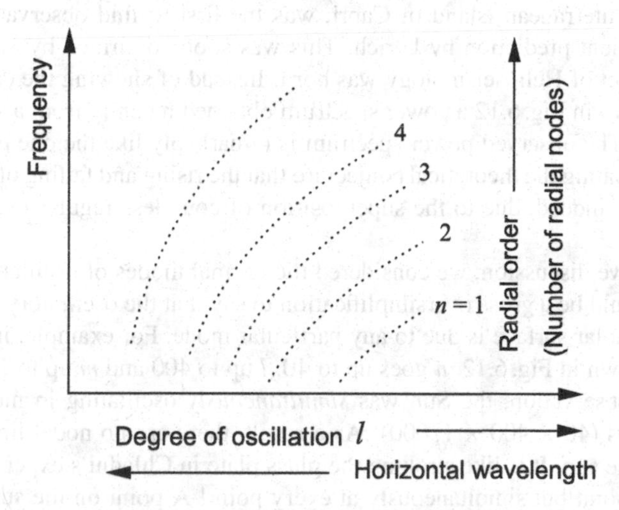

Fig. 6.11 The theoretically predicted frequency spectrum of an oscillating gas sphere. Plotted on the X- axis is the degree of oscillation, l (also known as the Spherical Harmonic Degree). The reciprocal of this is the horizontal wavelength of the sound waves. The various dotted curves correspond to the various radial orders (the number of nodes in the radial direction). For a given n and l, there are $(2l + 1)$ modes with different azimuthal number m. But *for a symmetric and nonrotating sphere, all these* $(2l + 1)$ *modes will have the same frequency*

Let us now fix l, and go to modes with greater number of radial modes. It should not be difficult to convince yourself that, again, the frequency of the modes will increase as n increases. Hence, the frequency spectrum of a vibrating sphere will look like a series of strings of pearls, stacked on top of each other (Fig. 6.11).

Let us now return to our discussion of the oscillation of the Sun's surface. We mentioned that Ulrich and Leibacher and Stein suggested that the surface oscillations observed by Leighton might be caused by sound waves trapped in the convection zone. They argued that the Sun acts as a resonant cavity, with sound waves known as *p modes* (or pressure oscillations) being trapped between the solar surface and the lower boundary of the convection zone. Ulrich went further: he argued that if the above suggestion was correct then the strongest solar oscillations will fall into a series of narrow bands when the amplitude of the oscillations are displayed in a two-dimensional diagram displaying the period (or frequency) versus the horizontal wavelength (or degree of oscillation). In other words, Ulrich argued that it would look like Fig. 6.11. Such a two-dimensional plot is known as the *power spectrum.*

Helioseismology

In 1975, Franz-Ludwig Deubner, a German astronomer who was observing the Sun from the Mediterranean island of Capri, was the first to find observational proof for this prescient prediction by Ulrich. This was soon confirmed by other groups, and the subject of Helioseismology was born. Instead of showing the early results, we have shown in Fig. 6.12 a power spectrum obtained recently from a space-borne observatory. The observed power spectrum is remarkably like the one predicted by Ulrich, vindicating the theoretical conjecture that the rising and falling of the surface of the Sun is, indeed, due to the superposition of countless regular oscillations of the Sun.

In the above discussion, we considered the normal modes of oscillation of a gas sphere. It would be a gross oversimplification to say that the oscillatory motion of a patch of the solar surface is due to any particular mode. For example, in the power spectrum shown in Fig. 6.12, n goes up to 40; l up to 400 and m up to 1,000. Thus, during this observation, the Sun was *simultaneously* oscillating in more than 10 *million* modes ($40 \times 400 \times 1,000$)! As a result, there are no nodal lines or nodal surfaces in the sun. It is like stroking the glass plate in Chladni's experiment not at a particular point but simultaneously at every point! A point on the surface of the Sun may move up or down by as much as a metre or two, with a velocity of only a few centimetres per second due to any particular mode. Not all the modes will be in phase. As a result of the superposition of various modes, the resultant velocity amplitude of a patch on the surface could be as large as a few hundred metres per second.

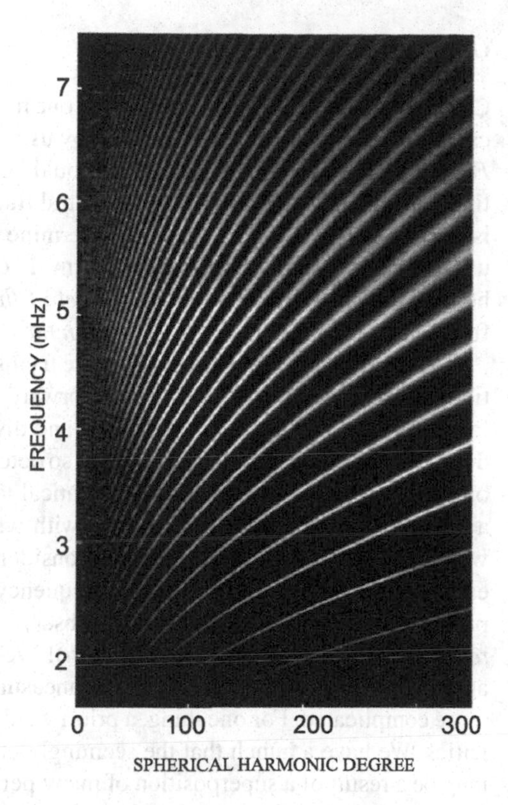

Fig. 6.12 The measured frequency spectrum of the Sun, obtained using the Michelson Doppler Imager on the SOHO spacecraft. This particular spectrum was derived from *uninterrupted* data recorded over 340 days! [Courtesy of SOHO/MDI consortium. SOHO is a project of international cooperation between ESA and NASA]

Let us remind ourselves of the primary objective of Helioseismology. By studying the slight motions of the surface, we wish to infer the conditions that obtain inside the Sun. There are two steps in this. First, we must decompose the observed motion into the normal modes of oscillations. Next, we must predict the expected amplitudes of the various modes and compare them with the observations. It is this second step that uses the temperature and density profiles from the standard model of the Sun.

The task of decomposing the observed motions into millions of normal modes might, at first sight, appear quite hopeless. But there is a well-developed branch of mathematics, known as *Fourier analysis*, to cope with precisely such a problem. This technique allows one to describe an arbitrary waveform as a superposition of sine and cosine waves. Fourier analysis could be used, for example, to describe the ripples on the surface of a swimming pool in terms of a superposition of sine and cosine waves. To take another example, when a signal generator produces a *square wave,* it is really adding up sine and cosine waves of different frequencies and different amplitudes. Fourier analysis is a bread-and-butter technique in a vast variety of problems in physics and engineering.

Observational Techniques

Coming back to the problem on hand, if one has data in the form of a *time series,* one can look for *periodic components* in it by using the well-known technique known as *Fourier transform.* Such a *time series* could be the intensity of light as a function of time, or wavelength of a particular spectral line as a function of time. The key thing is the accuracy with which one can determine the frequencies of the various modes using the technique of Fourier transform. If one wants to use Helioseismology to help us to constrain the *Standard Model of the Sun* then one has to determine the frequencies to better than *one part in ten thousand.* Herein lies the difficulty.

To illustrate this, imagine that you are trying to measure the frequency of oscillation of a pendulum. The most straightforward method is to count the number oscillations of the pendulum in a given time and divide the number of cycles by the time duration. The *accuracy* of the frequency so determined will depend upon the duration of the observation. To put it in mathematical terms, if we observe the pendulum for a time interval, Δt, then the precision with which we can determine the frequency will be proportional to $(\Delta t)^{-1}$. Let us consider the five-minute oscillation discussed earlier. If we want to determine the frequency of this mode to an accuracy of one part in ten thousand then we have to observe the Sun for 35 days (in other words, *ten thousand cycles of five minutes each*)! We have made this point mainly to draw attention to an important concept in the measurement of frequencies. Our problem is more complicated. For one thing, a priori we do not know if our data contain periodicities. We have a hunch that the seemingly chaotic phenomenon we are observing may be a result of a superposition of many periodic oscillations. We would first like to establish this, and, in addition, determine the frequencies as accurately as possible. As mentioned above, the Fourier transform of the data in the form of a time series will yield the *power spectrum.* If the data contain periodicities then they will show up as *peaks* in the power spectrum. The question is how well we can determine the frequencies of these peaks. The *frequency resolution* of the observation is directly determined by the reciprocal of the duration of the observation. Unfortunately, the Sun can be observed continuously only for a few hours. This severely limits the precision with which we can determine the frequencies in the power spectrum. There is a further complication. *Breaks* in the data (such as arising from the Sun setting at the observatory!) will introduce spurious frequencies in the power spectrum. These artificial frequencies (known in the trade as *alias* frequencies) will be the harmonics of the reciprocal of the duration of the observation.

The moral to be drawn from the above discussion is that for Helioseismology to provide useful constraints on the predictions of the Standard Model one must observe the oscillations of the solar surface for as long as possible.

The Antarctic

The Antarctic provides excellent opportunities to the solar astronomer since, weather permitting, one can observe the Sun continuously for six months. During the last three decades, many countries have set up ambitious scientific stations there. The first to exploit this were two French astronomers, Gerard Grec and Eric Fossat. For five continuous days in January 1980, they observed the Doppler shift of a spectral line from the sodium atom in the integrated light from the disc of the sun. This enabled them to detect, for the first time, meridional oscillation of very low degree ($l = 0,1,2,3$). This was a major breakthrough for two reasons. Since the light from the entire disc was used in this observation (as opposed to the light from a small patch of the Sun), it provided the first evidence that the oscillations were *globally coherent*. In other words, the entire Sun was participating in these coherent oscillations. Secondly, since these oscillations with large *horizontal wavelengths* penetrate deep into the solar interior (see Fig. 6.3) they provide the most crucial information.

Global Networks

Encouraged by the success of the Antarctic observations, astronomers rediscovered another way of ensuring that *the Sun never sets in the astronomer's empire* (a take-off by an astronomer on the famous Victorian saying, 'the Sun never sets over the British empire!'). They set up identical instruments at observatories located at different longitudes and combined the data from these instruments. One such network is known as GONG (Global Oscillation Network Group), managed by an international consortium of astronomers. This network consists of six stations located in California, Hawaii, Chile, Australia, India and the Canary islands, and sees the Sun 90 percent of the time (the GONG instrument in India is located in Udaipur in Rajasthan). The data from this network have yielded many exciting results, but we shall not npause to elaborate on this.

Solar Observatories in Space

The next step towards uninterrupted observation of the Sun was, of course, to set up solar observatories in outer space. This is, after all, the era of exploring the Universe from outer space. Of the many space-borne solar observatories, we shall single out one, shown in Fig. 6.13, named the Solar and Heliospheric Observatory (**SOHO**). This was launched in December 1995 by the ESA (European Space Agency) and NASA. This observatory houses twelve different instruments for a variety of observations of the Sun, with three of them being specific to Helioseismology. The most remarkable aspect of this observatory is that it has a truly uninterrupted view of the Sun. Being in space does not guarantee this since the Earth can eclipse the Sun as the satellite revolves around the Earth. To avoid this, SOHO has been placed in a special orbit at a point known as the *inner Lagrangian point* L1. This point is in the line joining the Sun and the earth, *at a distance of 1.5 million kilometres from the earth*. Figure 6.14 explains the significance of this point.

Fig. 6.13 The *Solar and Heliospheric Observatory* (SOHO). This was launched in December 1995 by ESA (European Space Agency) and NASA of USA. It weighs roughly 2,000 kg, and its dimensions are roughly 4 × 3 × 4 m. There are twelve separate instruments onboard this spacecraft. One of the unique features of this observatory is that it has an uninterrupted view of the Sun. SOHO moves around the Sun in step with the Earth, at a location known as the Lagrangian Point L1, where the combined gravity of the Earth and Sun keep SOHO in an orbit locked to the Earth–Sun line. The L1 point is approximately 1.5 million kilometres away from Earth (about four times the distance of the Moon from the Earth), in the direction of the Sun. There, SOHO enjoys an uninterrupted view of the Sun. All previous solar observatories have orbited the Earth, from where their observations were periodically interrupted whenever the Earth eclipsed the Sun

Let us first go to a frame of reference co-rotating with the Sun–Earth binary system. In this frame, let us draw the gravitational equipotential contours. The significance of these contours is that the *normal* to it gives the direction of the gravitational force. Clearly, close to the two bodies, these equipotential surfaces will be spheres centred on the two bodies; their projection on a plane would be circles. As we go farther and farther from the Earth, the gravitational force on a test particle due to the Sun will become more and more significant until, at the point L1, the equipotential contours intersect to form a horizontal *figure-of-eight.* This point of neutral equilibrium is known as the *inner Lagrangian point.* As we shall see in the third volume of this series, this point assumes special significance in the life history of binary stars. The point of interest to us here is that because SOHO has been placed at this point, it has an uninterrupted view of the Sun. Thanks, mainly to this vantage point, SOHO has made several monumental discoveries.

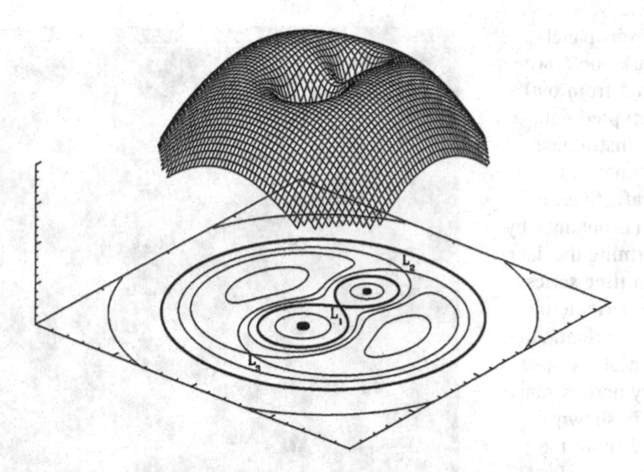

Fig. 6.14 At the top, is a three-dimensional representation of the gravitational potential in a binary system, such as the Sun–Earth system. In this particular figure, the ratio of the two masses has been assumed to be two. This surface, projected on to the plane below gives the gravitational equipotential contours; contours on which the potential is a constant. This representation is made in a frame co-rotating with the two stars. The figure-eight-shaped contour in the equipotential plot at the bottom are called the Roche lobes of each star. The points L1, L2 and L3 are known as the *Lagrangian points*, where the gravitational forces due to the two stars cancel out. If one of the stars becomes as large as its Roche lobe then mass can flow through the saddle point L1 from that star to its companion. Imagine that you are filling the three-dimensional potential hole with water. When one of the holes fills up, water will overflow to the adjacent hole. The SOHO spacecraft is placed at the point L1, from where it has an uninterrupted view of the Sun. [From Wikipedia, the Free Encyclopedia. Author: Marc van der Sluys, 2006. *Source* http://hemel.waarnemen.com/Informatie/Sterren/hoofdstuk6.html#h6.2]

The quality of data from this observatory is quite unprecedented. We have already seen an example of this in Fig. 6.12. That particular power spectrum was obtained from uninterrupted data obtained over a period of 340 days using an instrument known as MDI (Michelson Doppler Imager, named after the great experimentalist **Albert Michelson** whose measurements of the velocity of light laid the foundations for the *Special Theory of Relativity* proposed by Einstein in 1905).

In the top panel of Fig. 6.15, we have shown the Fourier transform of a time series stretching to 690 days (!) obtained with an instrument onboard SOHO known as GOLF (Global Oscillations at Low Frequency). The bottom panel shows an enlarged version of the power spectrum over a very narrow frequency range. With continuous data from 690 days, the frequency resolution is an impressive 17 *nanohertz.*

The Standard Model Put to Test

In 1926 Eddington asked the following question: 'What appliance can pierce through the outer layers of a star and test the conditions within?' Well, we now have the tools

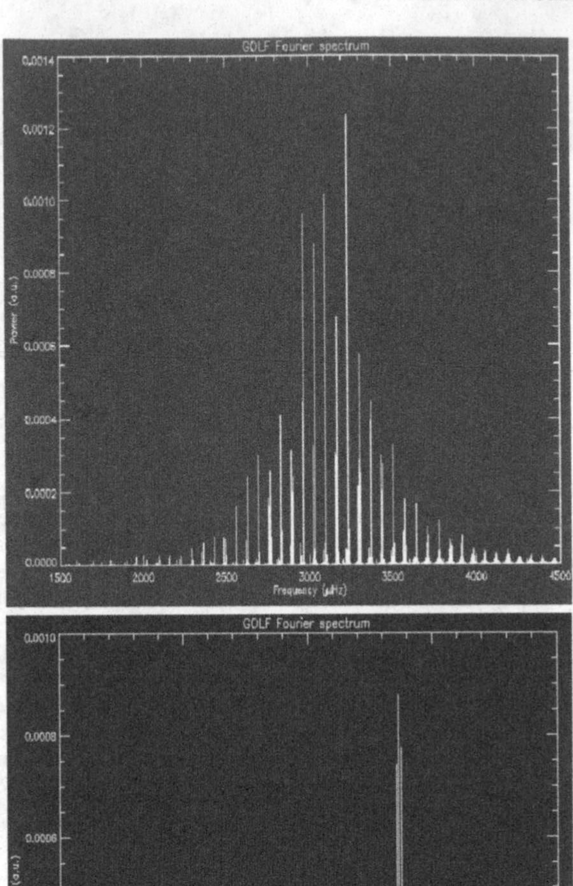

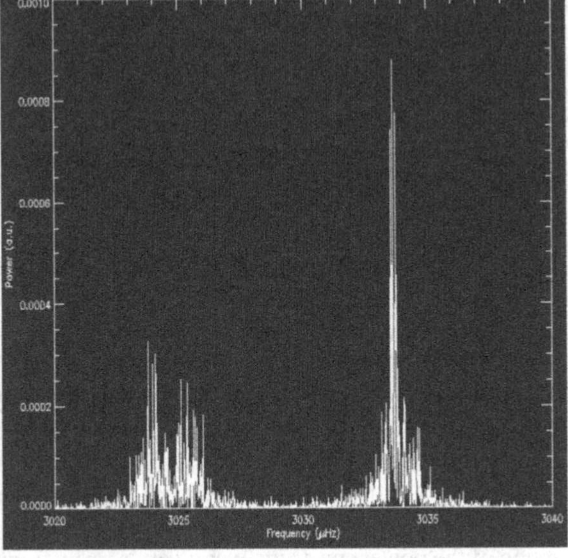

Fig. 6.15 The top panel shows the remarkable *Power Spectrum* derived from 690 days of uninterrupted data obtained by an instrument called GOLF onboard the SOHO spacecraft. Such a power spectrum is obtained by Fourier transforming the data in the form of a time series, and reveals the periodicities (or frequencies) present. An enlarged version of the spectrum over a very narrow range of frequencies is shown in the *lower panel* (note the scales in the *two panels*). Thanks to continuous data over 690 days, the frequency resolution (or the accuracy with which the frequencies can be determined) is *seventeen nanohertz*! [Courtesy of SOHO/GOLF consortium. SOHO is a project of international cooperation between ESA and NASA]

that Eddington would have loved to have—the frequencies of sound waves trapped inside the Sun. And we know them to an accuracy of better than one part in ten thousand. A detailed knowledge of these frequencies can be put to use in two ways.

We can use the temperature and density profiles given by the Standard Model, in conjunction with a theory of how the sound waves are trapped, to predict the oscillation frequencies of the normal modes and their amplitudes, and compare them with observations. If there is a discrepancy then we can tune the Standard Model till an agreement is obtained. This is known as the *forward method*.

Alternately, one can deduce one of the internal properties, say, the speed of sound as a function of depth, purely from the observations, and compare it with the sound speed profile predicted by the Standard Model. This is known as the *inverse method.* Such an inversion technique is employed in many fields like geophysics, atmospheric science, radiative transfer and so on. In all these problems, the basic idea underlying the inverse method is more or less the same. Unfortunately, it is rather technical and beyond the scope of this book. But for those of you who have studied some advanced mathematics, the principle of the inverse method may be described as follows. The observed quantity is related to the internal properties of the medium by an *integral equation*. That is, an integral over the path length involving an operator known as the *kernel* and some important internal property. In the present case, for example, the observed frequencies are integral measures of the sound speed along the path of the sound wave. To solve for the sound speed as a function of depth, one has to *invert* this nonlinear integral equation. This is a tough problem to crack. Therefore one resorts to some approximations that would enable one to simplify the problem. Traditionally, what one does is to *linearize* the integral equation: that is approximate the exact nonlinear equation by retaining only those terms that are linear. Such an approximation scheme is used very widely in physics.

In the problem on our hands, this simplification is introduced thus. As already remarked, the path of the wave and the depth to which it penetrates depends upon the radial order n and the degree l; the smaller the value of n or l, the deeper the wave penetrates before being refracted upwards once again. Since the frequencies of two waves moving along two different trajectories are different, the sound speeds along the two paths will also be slightly different. Therefore, the *difference in frequency* between two modes with slightly different degree or radial order can be used as a probe of the internal properties such as the *local temperature, local chemical composition, local motions* etc. that determine the local sound speed. Imagine modelling the Sun as consisting of thin concentric shells like an onion. By using the technique mentioned above, one can systematically derive the sound speed in each shell. I realize that all this is probably too technical for some of you. But do not worry if you could not make head or tail out of it. The above discussion was merely to convince you that there are well defined prescriptions for 'inverting' the problem and deriving the internal properties. This is, in principle, very similar to the way a CT-scan machine is used to generate a three-dimensional image of the human body.

Many groups around the world have succeeded in inverting such data, often combining data from several instruments. In Fig. 6.16 we have shown a comparison of the sound speed predicted by the Standard Model and the values inferred by inverting the observed frequencies obtained from the space borne observatory SOHO. Plotted along the x-axis is the radial distance from the centre to the surface measured in units of the radius of the Sun. Plotted along the y-axis is the normalized *difference* between the square of the predicted and derived sound speeds. Since the square of the sound speed is proportional to the temperature (remember Newton's formula for the speed of sound), this plot may be viewed as a comparison between the temperature

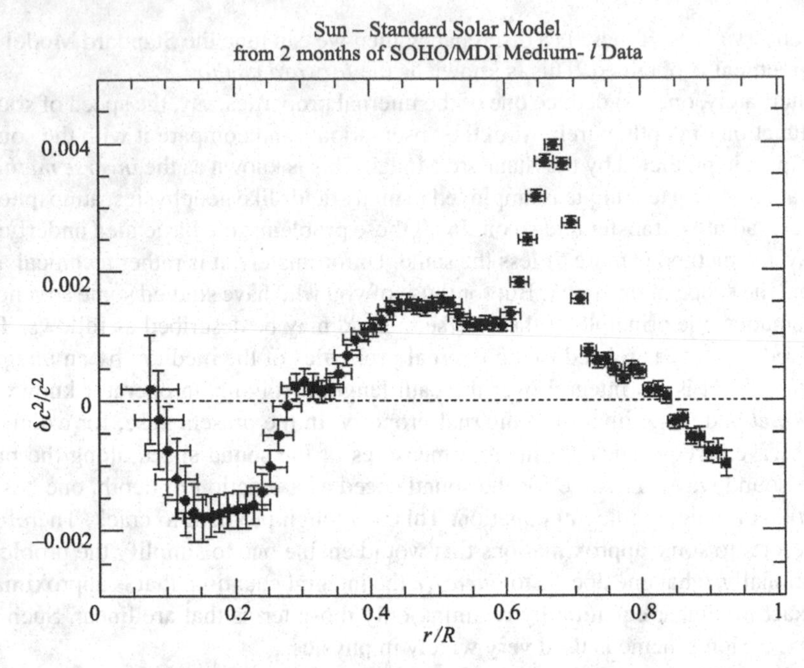

Fig. 6.16 This figure shows a comparison between the derived speed of sound inside the Sun and that predicted by the Standard Model of the Sun. What is plotted on the y-axis is the *difference* between the square of the derived sound speed and the square of the sound speed predicted by the Standard Model. As is customary, this difference has been made dimensionless by dividing by the square of the predicted sound speed. If there is perfect agreement between the two, then the value should be *zero*. The x-axis is the radial distance from the centre in units of the stellar radius. Notice that the maximum deviation from the predicted value is only 0.4 percent. And that occurs around $0.7R_{\odot}$ which is roughly where the base of the convection zone is. If we ignore this region—where special things may be happening—the maximum deviation is only 0.2 percent. Please note that the vertical *error bars* are much less than 0.2 percent. It is therefore safe to conclude that Helioseismology has proved that the predictions of the Standard Model of the Sun are accurate to within 0.2 percent. [Courtesy of SOHO/MDI consortium. SOHO is a project of international cooperation between ESA and NASA]

profile predicted by the Standard Model and that derived from observations. It will be seen from the figure that the maximum discrepancy is only about 0.4 percent. And this occurs just below the base of the convection zone at about 0.7 $R_{\odot}$. In the region where the energy generation takes place, namely, the *core* of the Sun (<0.2 $R_{\odot}$), the discrepancy is less than 0.2 percent. This is truly remarkable. But before we get too excited, we should be convinced that the errors involved deriving this plot are sufficiently small to warrant the above conclusion. Fortunately, they are. If you look at the *dots* in the figure, you will notice a vertical bar, as well as a horizontal bar, attached to them. The vertical bars indicate the errors in the results based on the errors in the determination of the oscillation frequencies. The horizontal bars provide a measure of the resolution in the inversion process. Clearly, the errors are small enough to warrant the conclusion that the temperature profile predicted using the Standard Model is accurate to better than 0.2 percent.

Let us recall the main objective of this chapter. In the Chap. 5 we discussed the pioneering experiment by Davis and his colleagues to detect the neutrino flux from the Sun. We saw that rather than jumping with joy at having finally detected the elusive neutrinos from the Sun, Davis and Bahcall were bothered (indeed, obsessed!) by the *discrepancy*—a factor of three between the predicted and observed flux of solar neutrinos. In 1968, when the first result from the chlorine detector was announced, it appeared that Bahcall was reading too much into this apparent discrepancy. Is the Standard Model of the Sun accurate enough for us to take the *discrepancy* seriously?

The solar neutrino flux is extremely sensitive to the temperature where the energy generation occurs. As we discussed in the Chap. 5, the probability of the fusion reactions (that produce the neutrinos) is very sensitive to the temperature. The quantum mechanical tunnelling probability which determines the reaction rate, and, therefore, the rate of neutrino production, is proportional to T^{25}. An error of 1 percent in the temperature corresponds to approximately 30 percent error in the predicted neutrino flux. An error of 3 percent in the temperature will lead to an error of factor of two in the number of neutrinos. So, the question of whether the discrepancy in Davis's experiment is significant or not boils down to how accurate are the predictions of the Standard Model, in particular the temperature near the centre of the Sun. Listening to the sounds of the Sun has provided us with the answer to this question. *Given this spectacular agreement between the predictions of the Standard Model and astronomical observations, we conclude that the disagreement by a factor of two or three between the predicted neutrino flux in Davis's experiment and the observed flux must be real.* The reason for the discrepancy must lie in fundamental physics. We shall discuss this at length in the Chap. 7.

Rotation of the Sun

Before winding up this chapter, let us briefly discuss one of the most exciting byproducts of Helioseismology, namely the rotation of the interior of the Sun.

A Bit of History

The fact that the Sun rotates has been known for nearly four hundred years. Soon after the telescope was invented, a number of persons used it to look at the heavenly bodies. We are all familiar with Galileo Galilei discovering the moons of the planet Jupiter using his telescope. Equally important was the discovery of small dark spots on the disc of the Sun, which have come to be known as *Sun spots*. This discovery was made around the year 1611, independently by a number of astronomers: **Galileo Galilei** in Rome, and **Johannes Fabricius** and **Christoph Scheiner** in Germany. Interestingly, this was an unwelcome discovery at that time. A spot on the Sun was regarded as a blemish, and this was unacceptable philosophically. Nature must surely be perfect! To give you an idea of how strong such a prejudice was, Scheiner's religious superior, instead of applauding him, admonished him for making such a claim. He is supposed to have said:

> ...such a thing has never been mentioned by any ancient philosopher. I have read my Aristotle through from beginning to end more than once, and found nothing at all like this. So keep quiet about this absurd idea, and don't make a fool of yourself in public. Instead, you should convince yourself that it is simply some fault in your eye or your telescope that makes you think that you saw spots on the Sun.

Interestingly, sunspots must have been known for a very long time before this because they can be seen with the naked eye—at least, the spots which are bigger than about 50,000 kilometres across (Think about why only spots bigger than this size can be *resolved* with the naked eye). Such giant spots are quite common when the Sun is active. If you want to try and see them with the naked eye, then the best time (and a safe time!) to do this would be just before the Sun sets (preferably over the sea); the black spots will stand out against the deep red disc of the Sun.

Despite these prejudices, Scheiner, in Germany, and Galileo, in Italy, continued to systematically observe the sunspots. One of the things they noticed was that over a period of a few days, the dark spots slowly moved across the face of the Sun. In Fig. 6.17, a drawing made by Scheiner in 1627 has been reproduced. In this sketch, he has depicted the location, and the shape, of two spots observed over 13 consecutive days. The two spots appear to move from the eastern limb to the western limb of the Sun. Both Scheiner and Galileo noted that the shape of the spots appeared to be distorted near the limb: while they were nearly circular on most days, they were more oval in shape when near the two limbs. From this they drew two remarkable conclusions. First, the spots must be on the *surface of the Sun*. They could not be, for example, tiny planets orbiting the Sun! Second, the spots appear to move across the disc of the Sun because the *Sun was rotating about its axis!* This interpretation also explained the fact that the speed with which the spots moved across the disc was not constant; the motion was fastest when the spots were near the centre of the disc, and the apparent motion was slower when the spots were at the limb.

And thus it was established that Sun rotated about its axis once in about twenty-seven days.

The next important discovery concerning the rotation of the Sun was made by the English amateur astronomer **Richard Christopher Carrington** in the 1850s. Although his father, a very rich brewer, wanted him to study theology at Cambridge University, young Carrington was drawn to astronomy. When he inherited his father's wealth, he built himself an astronomical observatory! From his sustained observations of the Sun during the period 1853–1861, he came to the remarkable conclusion that the Sun does not rotate as a rigid body. Carrington discovered that near the solar equator the rotation period is about 25 days, whereas it is about 27 days near 30° latitude. Later observations have confirmed this early finding, and have found that near the poles the rotation period is nearly 30 days. This is most intriguing. Ever

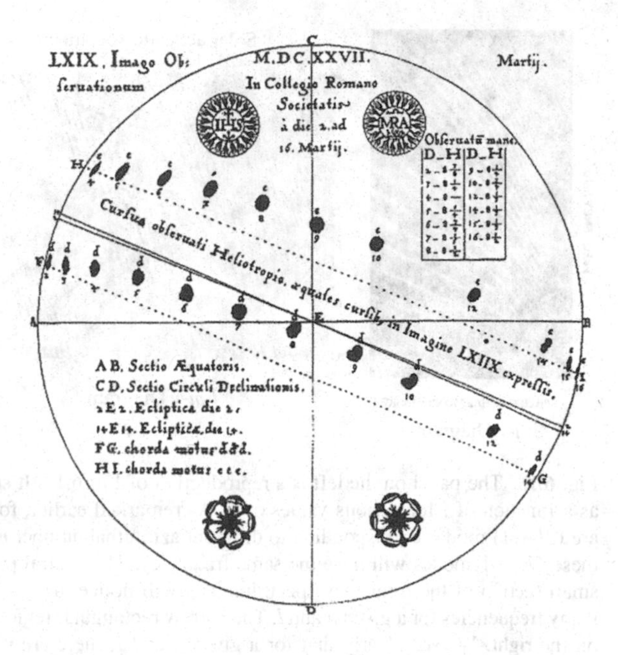

Fig. 6.17 This is a reproduction of a drawing made by Scheiner in 1627. Scheiner has carefully followed the position of two spots on a daily basis. They move from *left* to *right* in this drawing. From this he concluded that the Sun must be rotating about its own axis. [Drawings of sunspots from German mathematician Christoph Scheiner's *Rosa Ursina* 1630)]

since Carrington astronomers have been eager to find out how the interior of the Sun rotates.

Helioseismology Revisited

Helioseismology has provided a comprehensive answer to this question. Let us briefly recall our earlier discussion regarding the modes of oscillation of a sphere. We had argued that the various modes of oscillation can be characterized by three *quantum numbers*:

1. The *radial order*, n, specifying the number of radial nodes.
2. The *degree*, l, the number of nodes along the meridian.
3. The *azimuthal order*, m, the number of nodes along the latitude.

Further, we had remarked that if the Sun were spherically symmetric and *non-rotating* then all modes of a given n and l, but different m, will have the same frequency; recall that for a given l, m takes on $(2l + 1)$ values ranging from $-l$ to $+l$. In the technical jargon, one says that these $(2l + 1)$ modes are *degenerate*. We had argued that the frequency spectrum of a vibrating sphere will look like a series of strings of pearls, stacked on top of each other (see Figs. 6.11 and 6.12), with each *string* corresponding to a particular value of the radial order n but varying l. For any given l, modes with different m will have the same frequency. But if the star is rotating then this degeneracy of the $(2l + 1)$ modes is lifted, and the original frequency splits into

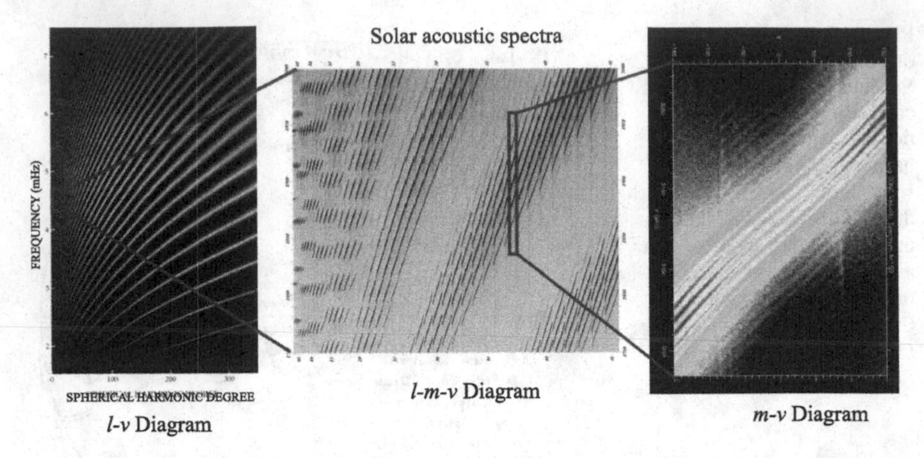

Fig. 6.18 The panel on the left is a reproduction of Fig. 6.12. It shows the measured frequencies as a function of l for various values of ν. As remarked earlier, for fixed values of n and l, there are $(2l + 1)$ modes corresponding to different azimuthal number m. If the star is *non-rotating*, all these $(2l + 1)$modes will have the same frequency. The central panel shows an enlargement of a small section of the frequency spectrum. You will notice that the frequency is, in fact, split into many frequencies for a given n and l. The narrow rectangular region is further enlarged in the panel on the right. We see clearly that for a given n and l, there are many frequencies. *This splitting occurs because the Sun is rotating.* [Courtesy of SOHO/MDI consortium. SOHO is a project of international cooperation between ESA and NASA]

$(2l + 1)$ frequencies. This may be seen in Fig. 6.18 in which the measured frequency for a given n and l has been magnified to show the splitting due to rotation. The magnitude of this splitting or, in other words, the difference in frequency between the $(2l + 1)$ modes, tells us about the rate of rotation of the stellar material in a shell in which the particular modes are trapped. Before discussing this further, let us digress to understand why such a splitting occurs. We shall first refresh our memory about a similar phenomenon in atomic physics.

Effect of a Magnetic Field on an Atom: Zeeman Effect

This splitting of a frequency into *multiplets* due to the rotation of the star should ring a familiar bell. It should remind you of *Zeeman Effect,* the splitting of the spectral lines emitted by atoms when a magnetic field is applied. Let us briefly recall this phenomenon. In 1896, **Zeeman** discovered that spectral lines in the light emitted by atoms are split up into components when the source emitting the lines is placed in a very strong magnetic field (Fig. 6.19). It might interest you to know that in 1862 **Michael Faraday** tried to investigate the effect of a magnetic field on the light emitted by a source. Unfortunately, he could not discover the effect that Zeeman discovered because his equipment did not have adequate resolution!

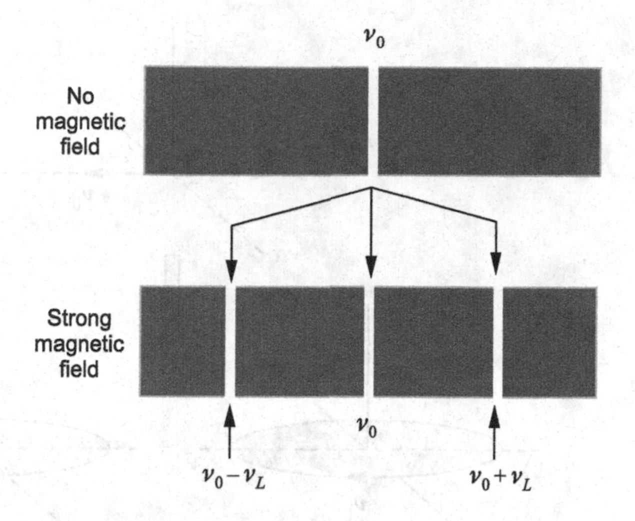

Fig. 6.19 Zeeman Effect: originally degenerate spectral lines are split into multiple components when a strong magnetic field is applied

Although a proper and satisfactory explanation of this had to wait for the advent of quantum mechanics, Zeeman was able to provide a simple explanation based upon the classical theory of matter that had just been developed by the great Dutch physicist Lorentz. According to this theory, radiation of a fixed frequency is emitted by electric charges when they execute simple harmonic motion; the frequency of radiation will be equal to the frequency of oscillation. Let us now discuss the effect of the applied magnetic field. It is useful to decompose the motion of the particle into a component parallel to the field and one perpendicular to the field. As you know, the magnetic field has no effect on the motion of charged particles moving parallel to it and hence the frequency of vibration of this component is unaffected. Let us call this frequency ν_0. The component of the motion *perpendicular* to the field will be affected. These particles will be deflected by the field (due to the $\vec{v} \times \vec{B}$ force) and will be forced to precess around the field with a frequency known as the *Larmor frequency* ($\nu_L = eB/4\pi mc$).

Now we are ready to answer the question we had posed earlier. The effect of an applied magnetic field on the radiation can easily be seen if it is first noted that a linear simple harmonic motion can be resolved into two superimposed circular motions, with the two motions being executed with the same frequency but in opposite directions. The magnetic field causes the frequency of these two circular motions to be different. It is a simple matter to show that to a very good approximation, these two frequencies will be $(\nu_0 + \nu_L) and (\nu_0 - \nu_L)$. To summarize, the effect of the magnetic field will be to split the original spectral line into three lines, with frequencies:

$$\nu_0$$
$$\nu_1 = \nu_0 + \nu_L$$
$$\nu_2 = \nu_0 - \nu_L$$

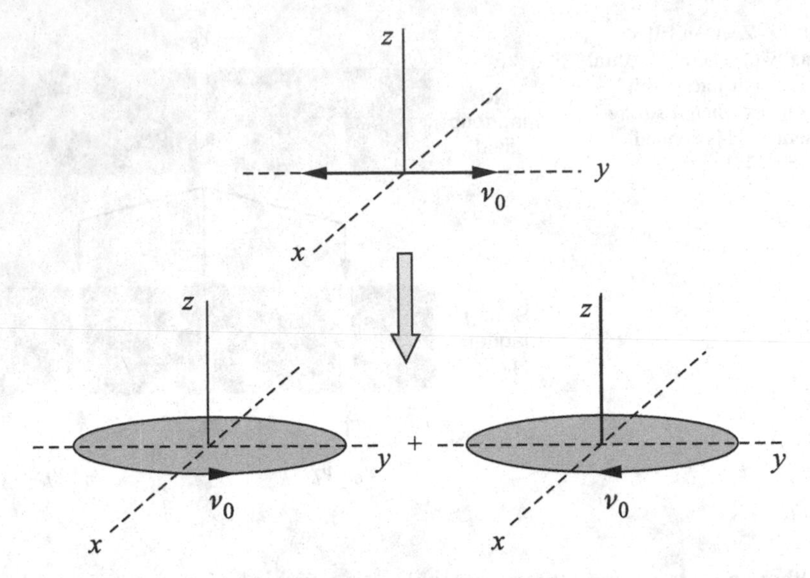

Fig. 6.20 Simple harmonic motion being executed by an electron can be represented as a superposition of two circular motions of the same frequency, but in opposite directions

This simple explanation by Zeeman was adequate to explain his original observations. Later and more-refined observations showed that there are many more components produced by the magnetic field. A proper explanation of these had to wait, of course, for the discovery of quantum mechanics. Nevertheless, the above explanation due to Zeeman captures the essence of the underlying physics.

Rotational Splitting of the Sound Wave Spectrum

Let us now attempt to understand along similar lines why the splitting of sound-wave frequencies occurs due to the rotation of the star. At first, recall that we are studying the spectrum of a three-dimensional standing wave pattern of sound waves. This pattern arises due to a constructive interference between a wave and its *twin* moving in the opposite direction. The important thing to appreciate is that sound waves are carried with the medium; a phenomenon known as advection (Fig. 6.20).

To an outside observer, the *prograde* wave (moving in the same direction as the sense of rotation of the local medium) will appear to complete one round a little faster than the *retrograde* wave (moving against the direction of rotation). Another way of saying this is that *frequency* of a wave, and its *twin*, are no longer the same; the frequency of one is slightly increased, while for the other member of the twin it is decreased (recall the discussion above of Zeeman splitting). It should be intuitively obvious that the frequency *shift* (or splitting) will depend upon the rate of rotation of the medium in the shell in which the wave under discussion is trapped, just as the

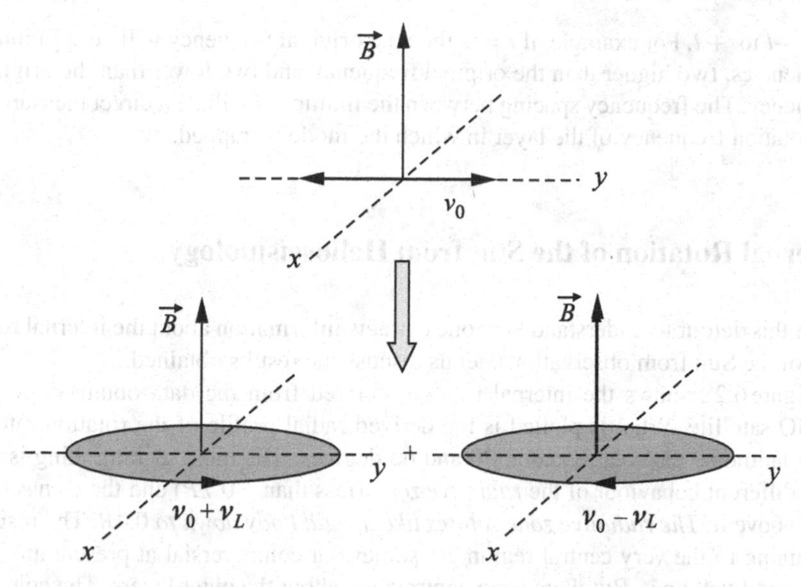

Fig. 6.21 When a magnetic field is applied along the z direction, the two circular motions are affected differently. If the original frequency is ν_0, then the frequencies of the two circular motions will now be $\nu_0 + \nu_L$ and $\nu_0 - \nu_L$, respectively

frequency splitting in Zeeman effect depended on the strength of the magnetic field (Fig. 6.21).

If you are more comfortable with a mathematical description, look at it this way. For a non-rotating star, modes with angular dependence $e^{i(\omega t \pm m\varphi)}$ are degenerate (that is, they have the same frequency). Let us assume that in the rotating frame slow rotation changes ω only slightly. Then in the inertial frame (in other words, for an observer outside the star), the *prograde mode* ($e^{-im\varphi}$) will be sped up and the *retrograde mode* ($e^{+im\varphi}$) slowed down. Let us state this mathematically. Let us denote the azimuthal coordinate φ in the inertial and rotating frame by φ_I and φ_R, respectively. We have $\varphi_I = \varphi_R + \Omega t$, where Ω is the angular velocity of rotation and t is the time coordinate. The phase of the mode has the form shown below:

$$\sigma t \pm m\varphi_I = \sigma_R t \pm m\varphi_R,$$

where the frequency σ measured in the inertial frame is given, in terms of the frequency σ_R measured by an observer in the star, by the equation:

$$\sigma = \sigma_R \pm m\Omega.$$

This is the result we were looking for. For an observer outside the star, the original frequency is seen to split into a *multiplet*. The number of frequencies in the multiplet will depend upon the degree of the mode since m takes on $(2l + 1)$ values, ranging from $-l$ to $+l$. For example, if $l = 2$ then the original frequency will be split into 5 frequencies, two higher than the original frequency and two lower than the original frequency. The frequency spacing between the multiplets will be a direct measure of the rotation frequency of the layer in which the mode is trapped.

Internal Rotation of the Sun from Helioseismology

After this detour to understand how one extracts information about the internal rotation of the Sun from observations, let us discuss the results obtained.

Figure 6.22 shows the internal rotation derived from the data obtained by the SOHO satellite. What is plotted is the derived radial profile of the rotation rate at three latitudes: the solar equator, 30 and 60 degrees. The most striking thing is the very different behaviour of the *radiative zone* (less than $\sim 0.7\,R$) and the *convection zone* above it. *The radiative zone rotates like a rigid body down to* $0.4R$. The results pertaining to the very central region are somewhat controversial at present and we shall not dwell on it. But there is no controversy about the outer layers. The convection zone behaves very differently from the radiative zone, and Richard Carrington would have been absolutely thrilled to see this! We see that the pattern of rotation at the surface, first deduced by Carrington, persists all the way down to the base of the convection zone. *The equatorial layer rotates fastest, and the rotation rate decreases as we go to higher latitudes.* To put it differently, the interior of the Sun rotates *differentially.* And in the convection zone, different layers slide over each other. Why this is so is one of the outstanding questions facing theorists. The observed behaviour does suggest that some mechanism exists for transporting angular momentum from high latitudes towards the equatorial plane (since the high-latitude zones rotate slower). The most popular candidate today for the mysterious agent responsible for the transport of angular momentum is *turbulence.* To transport angular momentum, there must be *friction* between adjacent layers. The familiar *molecular viscosity* is too small in gases to be effective. The only other viable alternative that physicists have been able to think of is turbulence which is known to introduce stresses in fluids. These stresses may mimic frictional forces. At least, that is the great hope in this and many other contemporary problems in astrophysics!

Although one may not understand the reason for this bizarre pattern of internal rotation in the Sun, there is something to cheer about. One of the great mysteries has been the origin of the magnetic field of the Sun. Quite clearly, there is a *dynamo* at work. The nature and the location of this dynamo have been the subject of much study. When the deep convection zone was discovered, many had conjectured that this dynamo may be located at the base of this zone. Now this seems more plausible. Refer to Fig. 6.22 showing the internal rotation profile. It seems very likely that there will be a great deal of *shear* at the interface between the radiative zone and the convective zone, since the adjacent layers are rotating differentially. And such a strong shear is just what the doctor ordered for producing large scale magnetic fields. Although the last words have not been said, one can say with some confidence

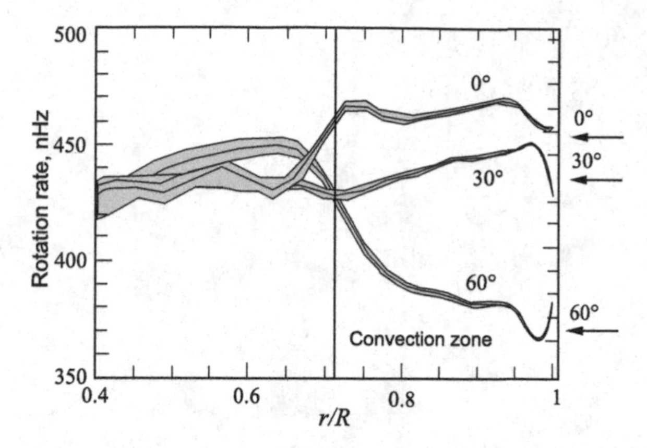

Fig. 6.22 The splitting of the frequencies in the solar acoustic spectrum has been used to derive the internal rotation rate of the Sun shown in this figure. The rotation rate at three different latitudes is shown. In the inner radiative zone, which is in hydrostatic equilibrium, the rotation rate is the same at all latitudes. This signifies *rigid rotation* of this region of the star. But in the convective zone, the Sun is rotating *differentially,* with the lower latitudes rotating faster. This is precisely what **Richard Carrington** had discovered in 1850 by looking at the sunspots. [Courtesy of SOHO/MDI consortium. SOHO is a project of international cooperation between ESA and NASA]

that unravelling the nature of the internal rotation represents a major breakthrough in solar physics.

It is time to wrap up this chapter. Let us recall the quotation from Eddington with which we began this chapter:

> At first sight it would seem that the deep interior of the Sun and stars is less accessible to scientific investigation than any other region of the universe. Our telescopes may probe farther and farther into the depths of space; but how can we ever obtain certain knowledge of that which is being hidden behind substantial barriers? What appliance can pierce through the outer layers of a star and test the conditions within?

Is it not remarkable that by listening to the sounds of the Sun we have been able to unravel so many secrets which were hidden inside its opaque interior! We have been able to deduce the internal temperature and density to an impressive precision.

So, we now know how the interior of the Sun rotates!

Chapter 7
The Smoking Gun is Finally Found

The Hunt for the Smoking Gun

In Chap. 5, 'Energy Generation in the Stars', we discussed the pioneering experiment by Raymond Davis and his colleagues to detect the neutrinos produced in the nuclear reactions that generate the energy radiated by the Sun. Indeed, the detection of these neutrinos was to be the ultimate test of the prescient conjecture by Eddington in 1920, and the detailed calculations by Hans Bethe in 1938. We saw that while Davis and his colleagues were ingenious enough to detect the neutrinos from the Sun, there was an apparent discrepancy between the predicted flux and the observed flux; the observed flux was roughly *one-third* of the predicted flux.

As you will recall, three classes of explanation were suggested to explain this.

1. The theoretical calculations were wrong. The prediction of the Standard Model of the Sun regarding the number of neutrinos produced per second was wrong, and/or the modelling of the interaction of the neutrinos in the detector and the consequent production rate of argon atoms in the detector was wrong.
2. Perhaps Davis' experiment was wrong.
3. Perhaps something happened to the original neutrinos as they travelled to the Earth.

All these possibilities were vigorously pursued in the decades that followed Davis' first results in 1968. As we saw in Chap. 6, a parallel development, namely, Helioseismology, gave us the tools to verify the Standard Model of the Sun. By 1997, astronomers were able to conclude that the temperature and density profile inside the Sun predicted by the Standard Model was accurate to about 0.1 precent. This left us with no choice but to accept that there is a discrepancy between the predicted neutrino flux and the observed flux provided, of course, the theoretical modelling of the neutrinos in the detector was correct *and* the efficiency of the chlorine detector was as good as Davis believed it to be. To be convinced of this, Bahcall and his colleagues put in an enormous amount of work to improve the modelling of the interaction of

G. Srinivasan, *What are the Stars?* Undergraduate Lecture Notes in Physics,
DOI: 10.1007/978-3-642-45302-1_7, © Springer-Verlag Berlin Heidelberg 2014

the neutrino with the detector. As we have already mentioned in Chap. 5, Davis and his colleagues spared no effort to improve their experiment.

But there was a nagging feeling that it would be good to have some more experiments—different kinds of experiments, perhaps—before one jumped to any conclusions. And so there were! The mystery was finally solved just a few years ago. As Bahcall would have put it, first the smoking gun was found, then the fingerprints on the gun and finally the culprits—the *missing neutrinos.*

This chapter is devoted to this remarkable detective story. But before we narrate it, let us refresh ourselves with some basic things concerning solar neutrinos.

Solar Neutrinos Revisited

In our discussion of the proton—proton chain reaction in Chap. 5, we saw that neutrinos were emitted in all the three branches of the chain. In essence, regardless of the branch, four protons are fused together to ultimately form a ^{4}He nucleus. The conservation of electric charge requires two positrons (e^+) to be created and the conservation of what is known as the *lepton number* requires two neutrinos to be created. The same is true of the CNO cycle also. Since these neutrinos are associated with electrons, they are called *electron neutrinos.* From now on, we shall be careful to attach a subscript to indicate the *flavour* of the neutrinos, namely the lepton with which they are associated.

Let us first recall the various reactions that produce these electron neutrinos and also the energy of the neutrinos produced in the various reactions.

$$
\begin{array}{lll}
{}^1\mathrm{H}+{}^1\mathrm{H} \rightarrow {}^2\mathrm{H}+\mathrm{e}^+ +\nu_e & (\text{p–p}) & E_\nu \leq 0.420\ \text{MeV} \\
{}^7\mathrm{Be}+\mathrm{e}^- \rightarrow {}^7\mathrm{Li}+\nu_e & (\text{p–p}) & (90\ \text{precent})\ 0.861\ \text{MeV} \\
 & & (10\ \text{precent})\ 0.383\ \text{MeV} \\
{}^8\mathrm{B} \rightarrow {}^8\mathrm{Be}+\mathrm{e}^+ + \nu_e & (\text{p–p}) & \leq 15\ \text{MeV} \\
{}^{13}\mathrm{N} \rightarrow {}^{13}\mathrm{C}+\mathrm{e}^+ +\nu_e & (\text{CNO}) & \leq 1.2\ \text{MeV} \\
{}^{15}\mathrm{O} \rightarrow {}^{15}\mathrm{N}+\mathrm{e}^+ +\nu_e & (\text{CNO}) & \leq 1.7\ \text{MeV}
\end{array}
$$

According to an ancient Chinese saying, one picture is worth ten thousand words. We have therefore reproduced in Fig. 7.1 the energy *spectrum* of the solar neutrinos predicted by the Standard Model of the Sun. This figure has been adapted from the spectrum given in the famous and definitive article by Bahcall et al. in *Reviews of Modern Physics* (1982).

The ^{37}Cl Experiment

Before describing some of the major experiments that followed the pioneering experiment by Davis, let us recall the salient features of that original experiment. The reaction that was used is the inverse of the laboratory decay of radioactive ^{37}Ar:

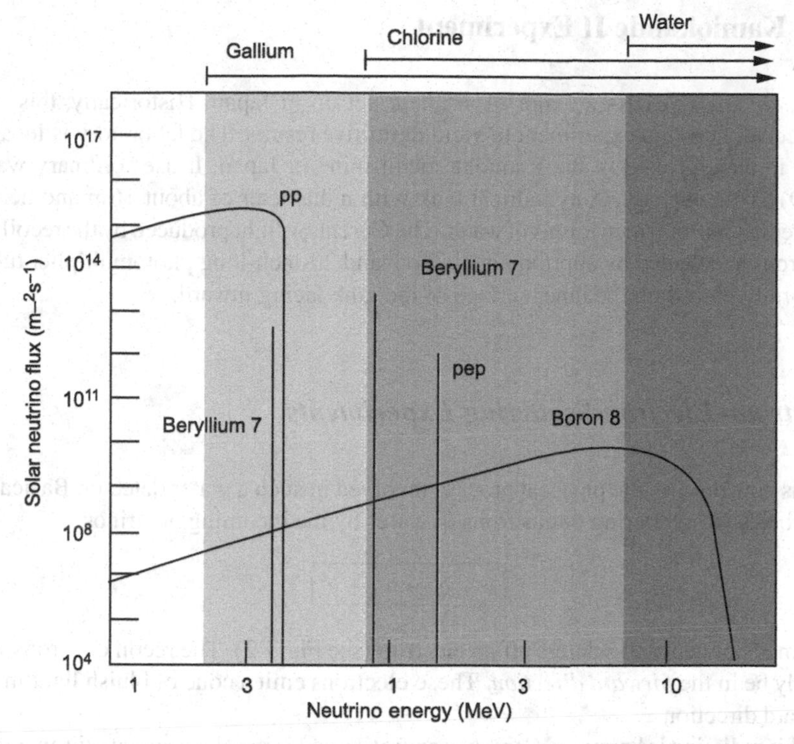

Fig. 7.1 Solar Neutrino Spectrum. The figure shows the energy spectrum of neutrinos predicted by the Standard Model of the Sun. The neutrino fluxes are given in units of number per m^2 per second per MeV at one Astronomical Unit (AU) from the Sun (1 AU is the average distance between the Sun and the Earth). This image has been taken from http://www.sno.phy.queensu.ca/sno/neutrino.html. *Courtesy* SNO

$$\nu_e + {}^{37}\text{Cl} \rightarrow e^- + {}^{37}\text{Ar}. \tag{7.1}$$

The threshold for this neutrino absorption is 0.814 MeV. This means that the neutrinos produced by the p–p reaction cannot be detected by this experiment. But the experiment will be sensitive to the neutrinos produced by the decay of ^{7}Be and the high-energy neutrinos produced by the decay of ^{8}B (Refer to the spectrum of the neutrinos shown in Fig. 7.1). Please also note that this experiment can detect only *electron neutrinos*. The event rate predicted by the Standard Model of the Sun was (7.9 ± 2.6) SNU; the uncertainty is indicated within parentheses. The observed rate in the experiment of Davis and his colleagues was (2.1 ± 0.9) SNU. As mentioned earlier, the results are expressed in terms of *Solar Neutrino Units* (SNU), which is the product of a characteristic calculated solar neutrino flux (cm^{-2} s^{-1}) times a theoretical cross-section for neutrino absorption (cm^2). Therefore, a SNU has the unit of *events per target atom per second* and is chosen for convenience to be equal to 10^{-36} s^{-1}.

The Kamiokande II Experiment

Next, we shall discuss a major experiment set up in Japan. Historically, this was the second neutrino experiment to yield definitive results. The laboratory is located 1000 m underground in the Kamioka metal mine in Japan. It uses ordinary water (H_2O) as the detector. A cylindrical tank with a diameter of about 16 m and height 16 m contains 3000 metric tons of water. The Cerenkov light produced by the recoiling electrons is detected by approximately thousand 20-inch-long photomultiplier tubes uniformly placed on the inner surface of the tank facing inward.

Neutrino–Electron Scattering Experiments

Let us now discuss the physical process involved in such a water detector. Basically, it involves the scattering of electrons in water by the incoming neutrinos:

$$\boxed{\nu + \mathrm{e} \rightarrow \nu' + \mathrm{e}'} \tag{7.2}$$

The incident neutrino scatters off an electron (see Fig. 7.2). The recoil electrons will mostly be in the *forward direction.* These electrons emit a cone of bluish light in the forward direction.

This radiation is known as *Cerenkov radiation* and is the electromagnetic analogue of *shock waves* created by a projectile moving faster than the speed of sound in the medium. Cerenkov radiation is emitted by particles moving with speeds greater than the phase velocity of light in the medium. (If you look down a *nuclear reactor* in which the uranium pile is immersed in water, you will see a bluish light; this is Cerenkov radiation).

1. An important feature of this water detector is that one will be able to reconstruct the direction from which the incident neutrino came. Since the electron is scattered in the *forward direction,* reconstruction of the electron tracks will give us a *vector* that points back in the direction from which the neutrino came.
2. Another very important feature is that neutrino—electron scattering will occur for neutrinos of any flavour, although it is much more sensitive to the electron neutrino. The scattering cross-section for the *electron neutrinos* (ν_e) is roughly 6.5 times *more* than for the *muon neutrinos* (ν_μ) and the *tau neutrinos* (ν_τ) put together.
3. Scattering experiments also provide the exact *arrival times* of the neutrinos. The moment when the photomultiplier tube detects a flash of Cerenkov light is essentially the moment when the neutrino interacts with the detector. It is this feature, together with being able to reconstruct the arrival direction that enabled the Japanese physicists to make a historic detection in 1987.

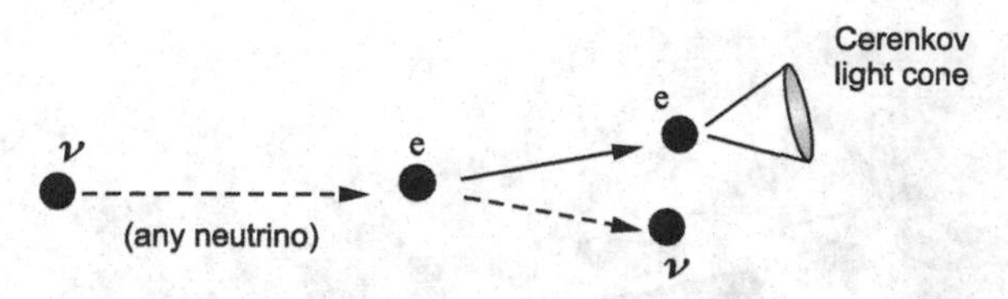

Fig. 7.2 The basic interaction in the water detector. The incoming neutrino scatters off an electron. As a consequence, the electron is accelerated in the forward direction, and emits a narrow cone of *bluish* light in the forward direction. This radiation is known as *Cerenkov radiation.* The direction from which the Cerenkov light comes tells us the direction from which the neutrino arrived. This process can occur for neutrinos of any of the three flavours

As we shall discuss in the next monograph in this series, in the standard scenario for the explosions of massive stars, known as *supernovae,* the core of the star collapses to form a very highly condensed quantum star known as a *neutron star*, which consists essentially of neutrons. During the collapse, almost all the protons are converted to neutrons through the following reaction: p+e→ n+ν_e. Since there are roughly 10^{57} protons in the collapsing core with a mass of about 1 solar mass, 10^{57} neutrinos would be produced during the birth of a neutron star. And since the collapse of the core of the star occurs in a few milliseconds, there should be an enormous burst of neutrinos accompanying the supernova explosion. This is the theoretical scenario for the most common type of supernovae. Although very plausible, this had remained a theoretical conjecture since the 1930s. The Kamiokande experiment was able to detect for the first time the burst of neutrinos from the *supernova* of 23 February 1987 in the Large Magellanic Cloud. The arrival time of the neutrinos, and their directionality, enabled one to associate this neutrino burst with the supernova in the Large Magellanic Cloud! In 2002, Masatoshi Koshiba was awarded the Nobel Prize for this discovery.

This water detector is very sensitive to the high-energy neutrinos. As was mentioned above, high-energy neutrinos are produced in the Sun only in the rare decay of the boron nucleus (see Fig. 7.1, showing the energy spectrum of the neutrinos). The original Davis experiment was also sensitive to these high-energy neutrinos.

In 1989, more than two decades after the first results from Davis' experiment, the Kamiokande II experimenters announced their finding. Just as in Davis' experiment with the chlorine detector, the number of solar neutrino events detected by Kamiokande II was *less* than predicted by the Standard Model of the Sun and the Standard Model of elementary particle physics. *But the discrepancy was less severe than observed with the chlorine detector.* The observed flux was ~45 precent of the theoretically predicted flux; recall that Davis detected ~33 precent of the predicted flux. This caused many eyebrows to be raised. It also provided the motivation for an even bigger water detector which would settle the issue once and for all! (Fig. 7.3).

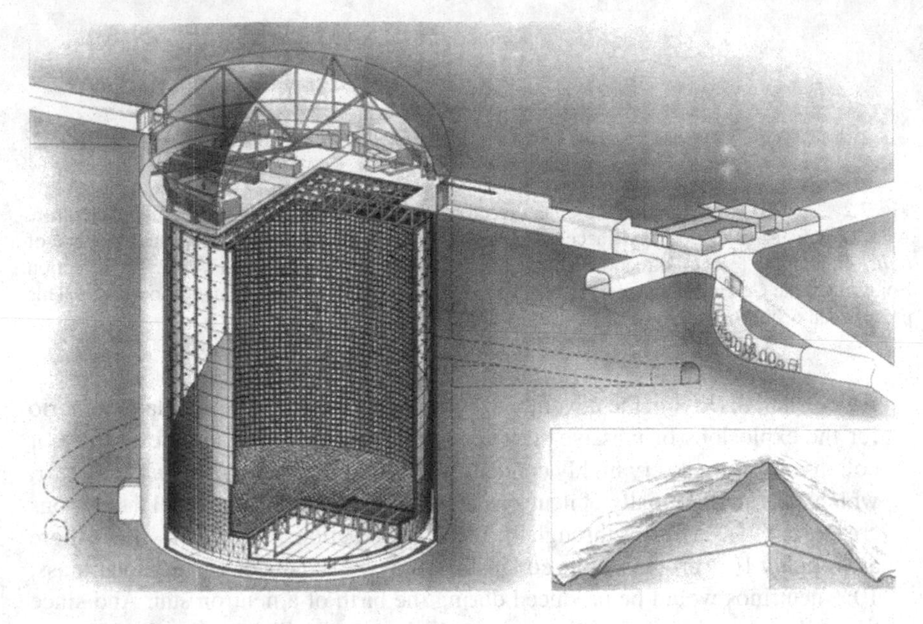

Fig. 7.3 The Super–Kamiokande Detector, University of Tokyo. The detector contains an inner volume filled with 32,000 tons of pure water, and an outer volume filled with 18,000 tons of pure water. The outer volume shields the inner volume from particles other than neutrinos. The inner volume is surrounded by 11,000 photomultiplier tubes that detect the Cerenkov light emitted by electrons accelerated by the neutrinos. (Drawing: Courtesy of Kamioka Observatory, ICRR, University of Tokyo)

The Super-Kamiokande Detector

Soon, a much larger version of the water detector was installed. This detector consisted of an inner volume containing 32,000 tons of pure water, and an outer volume containing 18,000 tons of water. The outer volume was intended to *shield* the inner volume from other kinds of *background events.* This inner volume of water was surrounded by 11,000 photomultiplier tubes which detected the Cerenkov light emitted by the recoiling electrons. This was a truly impressive experiment.

Within a few years after the Super–Kamiokande detector was commissioned, precise measurements of higher energy neutrinos by this new detector confirmed the magnitude of the deficit of the high-energy neutrinos found earlier by the Kamiokande II experiments. *There was a deficit of neutrinos, but this deficit was less severe than in the Davis experiment* (Fig. 7.4).

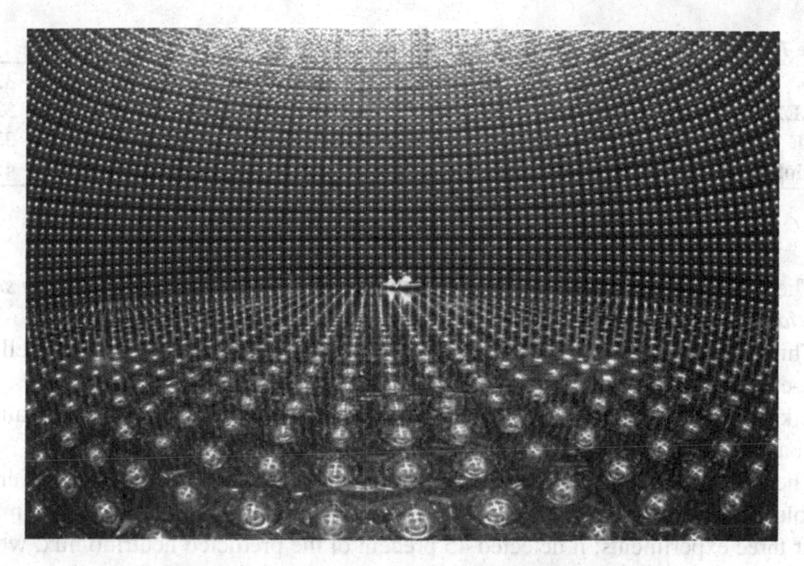

Fig. 7.4 A view of the inner volume partially filled with water. Two scientists may be seen in a boat, inspecting the photomultiplier tubes. (Photo: Courtesy of Kamioka Observatory, ICRR, University of Tokyo)

The Gallium Experiments

Independent of this, two other experiments were done in the 1990s, one in Italy and the other in Russia, using *gallium* as the detector. The *GALLEX* experiment used 30 tons of gallium in aqueous solution and was located in the Gran Sasso underground laboratory in Italy. The *SAGE* experiment (Soviet–American Gallium Experiment) used 60 tons of gallium metal as the detector underground in the high mountains in the Baskan Valley in the Caucasus Mountains. You should be impressed with the scale of these experiments because the total world production of gallium is only 10 tons per year! These experiments were particularly sensitive to the *low-energy neutrinos* from the p–p reaction which the chlorine experiment, as well as the water detector, could not detect (refer to Fig. 7.1). The gallium experiments were critical because theorists believed that they could calculate more accurately the expected flux of low-energy neutrinos ($E_\nu \leq 0.42\,\text{MeV}$). The reaction that was used in these experiments is given below.

$$\nu_e + {}^{71}\text{Ga} \rightarrow \text{e}^- + {}^{71}\text{Ge}, \qquad E_{\text{threshold}} = 0.2332\,\text{MeV} \tag{7.3}$$

The radioactive germanium produced in reaction decays by capturing an electron (the inverse of the above reaction with a lifetime of 11.43 days). As explicitly indicated in the reaction given above, these experiments, like the Davis experiment with chlorine, could only detect *electron neutrinos*.

Table 7.1 Conclusions of the four solar neutrino experiments

Davis	High energy ν	Only ν_e	Deficiency observed	Detected $\sim$ **33%**
GALEX	Low energy ν	Only ν_e	Deficiency observed	Detected $\sim$ **33%**
SAGE	Low energy ν	Only ν_e	Deficiency observed	Detected $\sim$ **33%**
Kamiokande	High energy ν	ν_e, ν_μ and ν_τ	Deficiency observed	Detected $\sim$ **45%**

The surprising result of these much-awaited gallium experiments was that *a substantial number of the lower-energy neutrinos were also missing.*

This deepened the mystery. It was now clear that both low-energy, as well as high-energy neutrinos were missing, *although not in the same proportion.*

Like a good detective does, it is useful to gather together the evidence obtained from all four experiments. This has been done in Table 7.1.

The most outstanding thing revealed by Table 7.1 is the following. The deficiency of solar neutrinos in the Kamiokande experiments is substantially less than in the other three experiments; it detected 45 precent of the predicted neutrino flux, while the other three experiments detected only 33 precent. *Could it be that the deficiency in the Kamiokande water detector is less than in other experiments because not all the neutrinos from the Sun were electron neutrinos? Could it be that a fraction of them were, in fact, muon neutrinos and tau neutrinos*? Kamiokande would have been able to detect the neutrinos of other flavours as well, whereas the other three experiments were capable of detecting only the electron neutrinos. But wait! The nuclear reactions in the Sun produce only electron neutrinos!

All this strengthened the suspicion that something must be happening to the neutrinos as they travel to the Earth from the centre of the Sun. In Chap. 5, we have already referred to the remarkable conjecture by Bruno Pontecorvo that neutrinos may change their identity back and forth between various *flavours* (the electron neutrino, ν_e, the muon neutrino, ν_μ, and the tau neutrino, ν_τ). If this happens, then, although only electron neutrinos are produced in the Sun, there will be an admixture of neutrinos of different flavours when they arrive at the Earth after travelling a long distance. Neither Davis' experiment nor the two Gallium experiments will be able to detect all the flavours, but Kamiokande would be! Strong and independent evidence that this might, indeed, be happening came in 1998 from entirely different quarters; from another beautiful experiment done with the Super-Kamiokande detector.

The Atmospheric Neutrinos

This experiment was designed to observe *muon neutrinos* produced in the Earth's upper atmosphere by cosmic rays; these are known as *atmospheric neutrinos.* Cosmic rays are extremely energetic particles that constantly bombard the Earth. A fraction of these originate from discrete sources within our own Milky Way Galaxy, while others come from distant galaxies. When these extremely energetic particles, travelling very

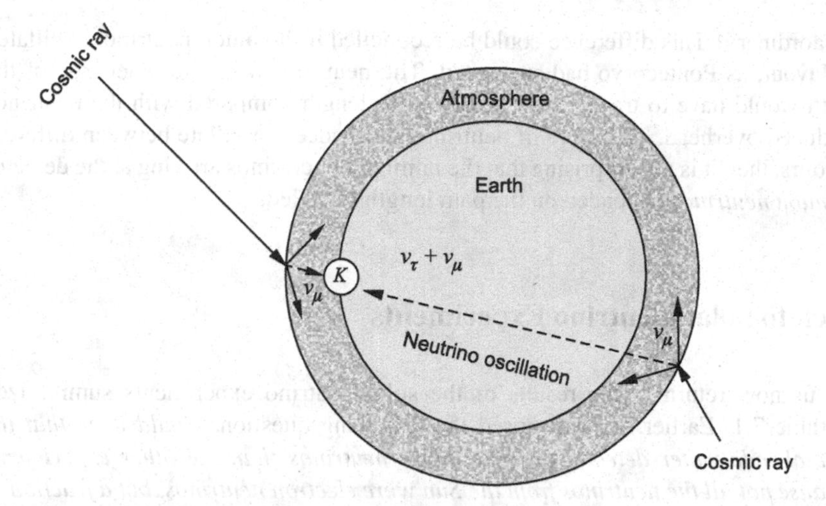

Fig. 7.5 The Kamiokande atmospheric neutrino experiment. Muon neutrinos produced in the upper atmosphere reach the detector from two different directions shown in the figure. The experiment found a statistically significant difference in the number of muon neutrinos arriving from the two directions. Since the path lengths are different, this result was consistent with the hypothesis of neutrino oscillations. Indeed, the observed difference in the muon neutrino flux agreed well with the theoretical predictions based on the hypothesis of neutrino oscillations

nearly at the speed of light, collide with atomic nuclei in our atmosphere a shower of *secondary particles* are produced. In some collisions, *muon neutrinos* are produced, and the objective of this experiment was to detect them. The first ever detection of atmospheric neutrinos was made in 1964 by a group of scientists from the Tata Institute of Fundamental Research, Mumbai. The experiment was located 2.3 km below the surface in the Kolar Gold Mines, near Bangalore. This was followed by another detection in South Africa.

Let us now return to the Kamiokande experiment which was designed to detect *muon neutrinos* produced in the Earth's upper atmosphere. Remember that the Kamiokande water detector was sensitive to neutrinos of all flavour. It was also vastly more sensitive than earlier experiments. What the experiment observed was this: The number of muon neutrinos that were detected depended upon *the direction from which they came.* Some of the neutrinos could have been produced in atmospheric events directly overhead of the detector. These neutrinos had to travel only 10–100 km of the atmosphere to reach the detector. On the other hand, the muon neutrinos could have also been produced on the other side of the Earth. Since the Earth is essentially transparent to the neutrinos they will have no difficulty in passing through the Earth. Some of these neutrinos will also be detected by the experiment. But these neutrinos would have travelled a larger distance (see Fig. 7.5).

What was observed in the experiments was this. The number of neutrinos arriving from the *overhead direction* was significantly different from the number that came from the other side of the Earth. Remember that the water detector gives us information about the direction from which the neutrino came (refer to Fig. 7.2). This is extraordinary! This difference could be reconciled if the muon neutrinos oscillated in flavour, as Pontecorvo had suggested. The neutrinos from the other side of the Earth would have to travel much greater path length compared with the neutrinos produced overhead. Therefore, if neutrinos did, indeed, oscillate between different flavours, then it is not surprising that the number of neutrinos arriving at the detector *as muon neutrinos* depended on the path length travelled.

Back to Solar Neutrino Experiments

Let us now return to the results of the solar neutrino experiments summarized in Table 7.1. Earlier, we had posed the following question: *Could it be that the Kamiokande water detector detected more neutrinos than the other experiments because not all the neutrinos from the Sun were electron neutrinos, but a fraction of them were, in fact, muon neutrinos and tau neutrinos*? Could a fraction of the original electron neutrinos have changed their identity to muon and tau neutrinos while in transit to the Earth from the Sun? The result of the atmospheric neutrino experiment, described above, lends very strong support to the idea that *neutrino oscillation* is the culprit responsible for the observed deficit.

Is there a way to clinch this argument? In the atmospheric neutrino experiment, the path lengths of the neutrinos coming from the two opposite directions were significantly different to enable us to draw a definite conclusion about the reality of neutrino oscillations. In the case of solar neutrinos, it is difficult to perform two experiments where the *distance* between the Sun and the detector is significantly different. To test the neutrino oscillation hypothesis, one must build a detector that can operate in *several modes*. In one of the modes, it must be able to accurately measure the flux of electron neutrinos alone. In an additional mode, it should be able to measure the combined flux of neutrinos of all flavours. This would settle the controversy once and for all. This is precisely what the Sudbury Neutrino Observatory was designed to do.

The Sudbury Neutrino Observatory

The SNO used the same basic principle as the Kamiokande experiments, namely, looking for Cerenkov light produced by the interaction of neutrinos with water. It is located 2000 m beneath the surface in a nickel mine in Sudbury, Ontario in Canada. This is the deepest underground experiment so far. Because it is located 2000 m below the surface, the cosmic ray background—which can also trigger the detector—is rather low. Only about three cosmic ray particles pass through the detector every hour. Like the Super–Kamiokande, this detector also has two volumes. The inner volume is a sphere 12 m in diameter, made of thick transparent acrylic, and holds 1000 metric tons of *heavy water* (D_2O). In heavy water, the molecules have two deuterium atoms instead of two hydrogen atoms. Recall that the deuterium nucleus consists of one proton and one neutron. As we shall presently discuss, this is the novel feature of this definitive experiment.

Surrounding this acrylic sphere is a stainless steel geodesic sphere of 18 m diameter on which are mounted 9500 sensitive photomultiplier tubes, each capable of registering single photons of the Cerenkov light. This sensitivity is needed because each neutrino event produces only 300–500 photons. This whole assembly is suspended in a cavern 22 m wide and 34 m high, carved out of solid rock and filled with 7000 tons of ordinary water.

Before narrating how this experiment finally solved the mystery of the *missing solar neutrinos* let us pause to appreciate the painstaking efforts made to discern the real neutrino events from various types of *background events*. In all great experiments in physics, it was the detailed understanding of the *errors* in the experiment that finally led to their success. It was the same in this experiment. We have already remarked that going two kilometres below the surface helped to drastically reduce the background events due to cosmic rays. But cosmic rays are not the only headache. Almost any material on Earth—steel, dust, even water—has a tiny amount of radioactive material.

These radioactive contaminants emit charged particles that can generate the same kind of Cerenkov light as neutrino interactions are expected to do. So every component of the experiment was *designer-made;* they were made out of material low in radioactivity. That was not all. The nickel mine chosen for the experiment was a functioning mine. Naturally, this made the logistics of running a laboratory a lot easier; all the access facilities were already there. But there was a heavy price to pay. Mine dust usually has high level of radioactivity. As one of the principal investigators of this experiment remarked, '*even a tablespoon of this dust dropped into the 275,000 gallons of heavy water would have enough radioactivity in it to mask the neutrino signals*'. So the entire underground laboratory was operated as a *clean room,* with the air continuously filtered. Every person entering the laboratory had to follow strict cleanliness procedures. This is the kind of clean environment demanded of a semiconductor laboratory in which computer *chips* are made. But those are tiny rooms. Imagine demanding a comparable level of cleanliness in an underground lab, 2 km beneath the surface! But that is the kind of precaution that had to be taken. As for the radioactive decay from the rocks, the 7000 tons of ordinary water in the outer cavern shielded the heavy water detector inside the acrylic sphere (Fig. 7.6).

Finally, where did they get 1000 metric tons of heavy water? One cannot go to a supermarket and buy heavy water! It so happens that heavy water is used as a *moderator* in some nuclear reactors; the moderator slows down the neutrons emitted when the uranium nuclei break up, thus making it possible for other nuclei to absorb them. For the Sudbury Neutrino Observatory, 1000 tons of heavy water was borrowed from the Canadian Nuclear Reactor Programme (Fig. 7.7).

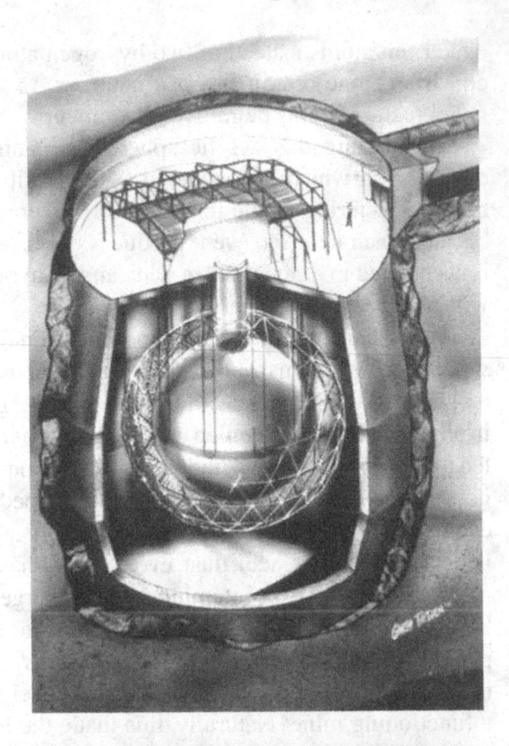

Fig. 7.6 Artist's drawing showing a cutaway of the Sudbury Neutrino Observatory detector. The inner sphere contains 1,000 tons of *heavy water* and is surrounded by a stainless-steel structure on which approximately 10,000 photomultiplier tubes are mounted. The outer barrel-shaped cavity is filled with purified ordinary water. This provides support and acts as a shield against particles other than neutrinos reaching the inner detector. Copyright Garth Tietien, 1991. Courtesy of SNO

Fig. 7.7 A view of the SNO detector. Approximately 10,000 photomultiplier tubes are mounted on this stainless-steel structure. Inside this structure is a sphere 12 m in diameter, made of thick transparent acrylic, containing 1,000 metric tons of *heavy water* (D_2O). Photo courtesy of Ernest Orlando Lawrence, Berkeley National Laboratory. Courtesy of SNO

Neutrino Interactions in Heavy Water

Like the ordinary water detector which we discussed in the context of the Kamiokande experiment, this heavy water detector also is sensitive to the high-energy neutrinos. Estimates show that at this depth about *five million solar neutrinos pass through every square centimetre of the detector per second*. Of these, only about *five* neutrinos are expected to produce any signal in any given *day*! There are three possible channels through which these neutrinos can interact with heavy water.

Channel 1: Absorption of the neutrino by a deuterium nucleus (only ν_e)

An *electron neutrino* can be absorbed by the neutron inside a deuterium nucleus, transforming it into a proton. An electron is emitted in the process.

$$\nu_e + \mathrm{n} \rightarrow \mathrm{p} + \mathrm{e}^-$$

The electron emits Cerenkov radiation. Such absorption can only occur for an electron neutrino. Such absorption cannot occur in ordinary water because a proton cannot absorb an electron neutrino; it can only absorb an electron anti-neutrino. Recall that all the reactions in the Sun produce only electron neutrinos but not their antiparticle.

Channel 2: Neutrino–electron scattering (ν_e, ν_μ, ν_τ)

The solar neutrino knocks off an electron from one of the D_2O molecules.

$$\nu + \mathrm{e}^- \rightarrow \nu + \mathrm{e}^-$$

This is the same interaction as in the Kamiokande detector, discussed earlier. The important thing to appreciate is that one can distinguish these electrons from those produced when a neutrino is *absorbed* by a deuteron. In the present case, the electron will be scattered in the *forward* direction and will emit a cone of Cerenkov light in the forward direction. Since the electron is scattered in the forward direction, reconstruction of the electron tracks will give us a *vector* that points back in the direction from which the neutrino came. This vector should point to the Sun. But there is no such restriction in the neutrino absorption process described in Fig. 7.8.

This *scattering* process can be triggered by any type of neutrino (ν_e, ν_μ or ν_τ), but not with equal probability. *It happens approximately 6.5 times more often for electron neutrinos.*

Channel 3: Break-up of a deuteron by the neutrino (ν_e, ν_μ, ν_τ)

This third reaction is also sensitive to all flavours of neutrino (ν_e, ν_μ or ν_τ). The neutrino breaks up the deuteron into a neutron and a proton.

$$\nu + \mathrm{d} \rightarrow \mathrm{n} + \mathrm{p} + \nu$$

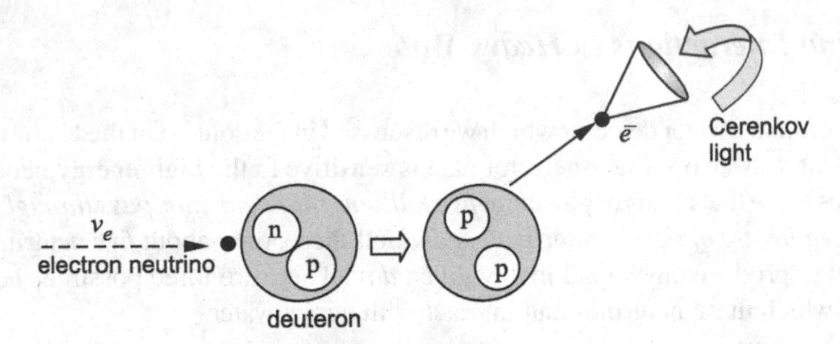

Fig. 7.8 Channel 1: The incident *electron neutrino* is absorbed by the neutron inside a deuterium nucleus, transforming it to a proton. An electron is created in the process. The Cerenkov radiation emitted by this fast-moving electron is detected. Note that the electron is not necessarily emitted in the same direction as the velocity vector of the incoming neutrino. Also, *such absorption is possible only for an electron neutrino*

The neutron thus released is soon captured by another deuteron, producing a gamma ray. This gamma ray scatters an electron in the heavy water, and it is this secondary electron that produces the Cerenkov light.

Not only can such break-up of the deuteron occur with any of the three flavours of neutrino, *they occur with the same probability*. That is, the electron neutrino, the muon neutrino and the tau neutrino all have the same footing as far as this reaction goes.

To summarize, with the *heavy water* detector one can make three independent measurements. Of these, neutrino absorption alone is sensitive only to electron neutrinos. The other two reactions (deuteron break-up and electron scattering) can be triggered by all the three types of neutrinos.

Now it is possible to directly test the hypothesis of neutrino oscillations. If the number of neutrino events measured with either electron scattering (channel 2) or deuteron break-up (channel 3) is *greater* than the number measured with the neutrino absorption reaction (channel 1), then it implies that neutrinos of flavour other than electron neutrinos (muon and/or tau neutrinos) must also be present in the flux from the Sun. But since the Sun produces only electron neutrinos, the transformation to other flavours must have happened on the way. Neutrinos must have oscillated between the three flavours during the transit (Fig. 7.10).

There already was an indication of this from the Kamiokande experiments, which we highlighted in Table 7.1. Recall that although these experiments (using electron scattering) recorded a deficit in the number of neutrinos, the deficit was *less* than in the chlorine experiment. The chlorine experiment had detected only about 33 percent of the predicted number. But the number of neutrinos detected in the Kamiokande experiments was about 45 percent of what had been predicted by theory. This could easily be reconciled if one postulated that the Kamiokande electron-scattering detector may have detected some muon and tau neutrinos, in addition to electron neutrinos. But this was a conjecture, since the Kamiokande detector could not separately detect only the electron neutrinos. The Sudbury experiment had the potential to settle this. That is why it was built.

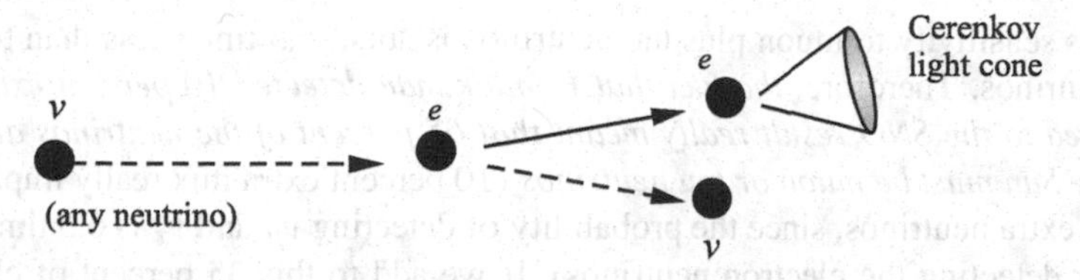

Fig. 7.9 **Channel 2**: This is the same as the reaction depicted in Fig. 7.2. The incoming neutrino scatters off an electron. As a consequence, the electron is accelerated in the forward direction, and emits a narrow cone of bluish light in the forward direction. This radiation is known as *Cerenkov radiation*. The direction from which the Cerenkov light comes tells us the direction from which the neutrino arrived. *This process can occur for neutrinos of any of the three flavours*

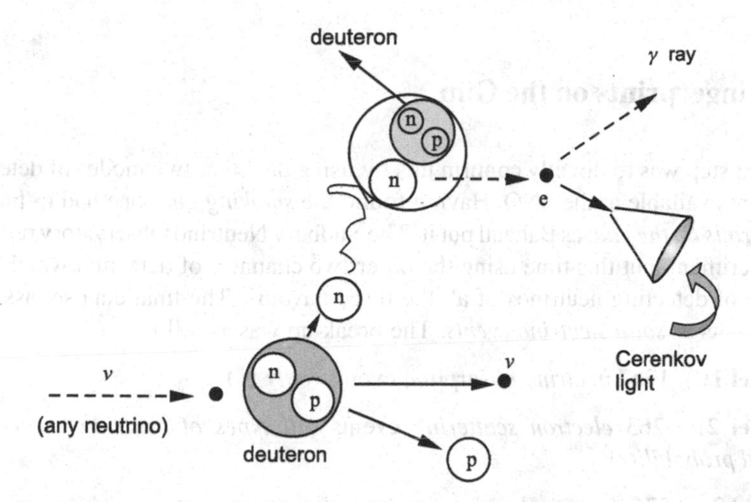

Fig. 7.10 **Channel 3**: Break-up of the deuteron. The incident neutrino can break up a deuteron into a free proton and a free neutron. The wandering neutron can be captured by another deuterium nucleus. A gamma ray will be emitted in the process. This gamma ray can scatter off an electron. The recoiling electron will emit Cerenkov radiation. *All three flavours of neutrino can trigger this process*

The Smoking Gun was Finally Found

To shed light on this, the Sudbury Neutrino Observatory first performed an experiment with the neutrino absorption reaction which is sensitive *exclusively to electron neutrinos* (Channel 1, shown in Fig. 7.8). After recording data for 241 days, SNO detected 950 neutrino absorption reactions. The Standard Model of the Sun had predicted that there should have been more than 2,700 neutrinos detected during this time. In other words, SNO *observed only 35 percent of the expected number*.

Suddenly the Kamiokande results made sense! Recall that Kamiokande had detected 45 percent of the expected number of neutrinos. The Kamiokande water detector was based on the neutrino–electron scattering reaction (Channel 2, shown in Fig. 7.9) which is sensitive to other types of neutrino also, although at a reduced level. Its sensitivity to muon plus tau neutrinos is about 6.5 times less than to electron neutrinos. Therefore, *the fact that Kamiokande detected 10 percent extra flux compared to the SNO result really meant that 65 percent of the neutrinos arriving from the Sun must be muon or tau neutrinos* (10 percent extra flux really implied 65 percent extra neutrinos, since the probability of detecting ν_μ and ν_τ is 6.5 times *less* than for detecting the electron neutrinos). If we add to this 35 percent of electron neutrinos, directly detected by SNO, then we have accounted for all the neutrinos expected from the Sun!

What followed was a historic announcement on 18 June 2001. The SNO collaboration announced that they had *solved* the solar neutrino puzzle. An ecstatic John Bahcall declared, 'the smoking gun has been found!'

The Fingerprints on the Gun

The next step was to directly confirm this by using the other two modes of detection that were available at the SNO. Having found the *smoking gun,* one had to find the *fingerprints on the gun*, as Bahcall put it. The Sudbury Neutrino Observatory repeated the experiment, but this time using the other two channels of detection which were capable of detecting neutrinos of all the three flavours. The final data set assigned 2806 events as *solar neutrino events.* The break-up was as follows:

Channel 1: 1967 *neutrino absorption* events (*only* ν_e).

Channel 2: 263 *electron scattering* events (*all types of neutrino events with unequal probability*).

Channel 3: 576 *deuteron break-up* events (*all types of neutrinos with equal probability*).

To translate these events to *number of neutrinos* arriving at the detector (per unit area per unit time) one has to take into account the *probability of detection* in each of the three channels (or reactions). When this was taken into account, the number of *events* mentioned above translated to number of solar neutrinos given below.

Channel 1: 1.75 *million electron neutrinos* pass through the detector per cm^2 per second. This was only 35 percent of the flux of neutrinos predicted. Importantly, this confirmed the earlier finding in the first run of the experiment.

Channel 3: In comparison, the 576 events assigned to deuteron break-up translated to a *total flux of 5.09 million neutrinos per cm^2 per second.* Recall that deuteron break-up can be triggered by neutrinos of all three types with *equal probability*. This was direct proof of the fact that the majority of neutrinos arriving from the Sun are either muon or tau neutrinos. It is important to stress that the difference between the total number of neutrinos detected (through Channel 3) and the number of electron neutrinos detected (through Channel 1) was determined with great accuracy. *The difference was more than five times the experimental uncertainty,thus making it a* *highly statistically significant discovery*. One could therefore say with considerable confidence that Sudbury Neutrino Observatory had actually detected neutrinos of all types. What is more, a total flux of *5.09 million neutrinos* per cm^2 per second was in remarkable agreement with the predictions of the Standard Model of the Sun.

Neutrinos do Oscillate in Flavour!

So the hostile witnesses to the transmutation of hydrogen into helium at the centre of the Sun had at last been found. All was well with astronomers' model of the Sun, and why it shines. But in confirming this, physicists have shot themselves in the foot. There is now compelling evidence that neutrinos do oscillate between various flavours. This, in turn, calls for a major revision of fundamental physics. Although this has nothing to do with the story of the stars, we must digress a little and try to understand the phenomenon of neutrino oscillations, at least qualitatively. After all, astronomy has, once again, provided a major input for basic physics.

As mentioned earlier in Chap. 3, neutrinos first entered the scene of the elementary particles in 1933. At that time, the list of elementary particles consisted of the proton, neutron and the electron; the neutrino was just a postulate. By the time the neutrinos were eventually discovered in 1956, the number of elementary particles had grown to more than fifty. In an attempt to bring some order, the known elementary particles were classified according to their mass: *leptons* (light particles), *mesons* (medium mass) and *baryons* (heavy particles). Now days, one uses the term *hadrons* to include the mesons and baryons.

When the terminology was first introduced, leptons were meant to be much lighter than the proton. Therefore, to include the electron, and the associated neutrino, in this family of leptons was appropriate. Today, the family of leptons includes some particles which are not so light—the muon and the tau. It also includes the associated neutrinos—the muon neutrino and the tau neutrino. It is these neutrinos that concern us. According to the *Standard Model* of elementary particles (known as the *Electro-Weak Unified Theory*), the leptons obey Fermi–Dirac statistics, and all the neutrinos have *zero mass*.

However, the evidence that we have been discussing of neutrinos changing their flavour requires that at least some of the neutrinos must have *non-zero mass*. Otherwise, as we shall now discuss, the phenomenon of neutrino oscillation cannot occur. Since neutrinos cannot have a mass in the Standard Model of elementary particles, the observed phenomenon of neutrino oscillations calls for a major revision of the Standard Model. But that is for the future.

The idea of neutrino oscillation was first put forward by Bruno Pontecorvo in 1957. At that time, Pontecorvo was not concerned with solar neutrinos. His idea was inspired by a similar phenomenon concerning the *K Mesons*. (I recommend that you look at *The Big and the Small* by G. Venkataraman for an excellent account of the fascinating story of elementary particles). Pontecorvo hypothesized that neutrinos might also oscillate between the various states,just as *K mesons* did. He was aware that for this to happen at least some of the neutrinos must have a nonzero mass, but this did not deter him. After all, nobody had proved that neutrinos had zero mass. One only knew that if at all they had a mass, it must be extremely small. Needless to say, no one took note of this radical suggestion by Pontecorvo. There was no compelling theoretical reason for entertaining this idea. Some ten years later, when Davis announced his famous solar neutrino deficit result in 1968, it was natural for Pontecorvo to point out that neutrino oscillations may explain this puzzle. More than forty years later, we have direct experimental evidence to confirm this remarkable conjecture.

The following discussion of neutrino oscillations is rather simple minded, but I believe it captures the essence of the underlying physics. If you have studied some elementary quantum mechanics, you will find it very straightforward to comprehend. If you are not familiar with the principles of quantum mechanics, two classical analogies should give you a feel for the phenomenon of quantum oscillations.

Mass States and Flavour States: Quantum Oscillations

The underlying principle of quantum physics is the duality between particles and waves. In quantum mechanics, the states of a system are described by what are known as *wave functions*. The basic thing to appreciate is the following. The neutrinos, or more correctly, the neutrino states $|\nu_e\rangle$, $|\nu_\mu\rangle$, $|\nu_\tau\rangle$ that are produced in weak interaction decays in association with the charged leptons e^-, μ^-, τ^-, respectively, are called *flavour states*. The point is that these flavour states are not states of definite mass (like the *electron* or *proton*, which have definite mass), but *linear combinations* of the more fundamental *mass states* (or mass *eigenstates*). The mass states are usually denoted by ν_1, ν_2 and ν_3. To put it differently, the flavour states are not normal modes of the system; the mass eigenstates are. *The mass eigenstates are the states in which the neutrinos propagate in vacuum.* And there are three mass states.

According to the rules of quantum mechanics, the wave function of, say, the *electron neutrino* is really a *linear superposition* of the wave functions of the three mass eigenstates. For the sake of illustration, let us assume for a moment that there are only *two mass eigenstates* and *two flavour states* (and not three.) In this case, the electron neutrino would be expressed as a linear superposition of the two mass eigenstates as follows: $\nu_e = \cos\theta\, \nu_1 + \sin\theta\, \nu_2$. Here θ is known as the *mixing angle.* Similarly, the muon neutrino can be expressed as a superposition of the mass eigenstates. When waves are *added*, one has to, of course, prescribe what the phase relationship between the waves should be. Therefore, what we call as the electron neutrino state is a linear superposition of the three mass states *with a particular phase relationship between the three waves that represent the mass states.* Similarly, the muon and tau neutrino states are obtained by adding the three mass states with different but specified phase relationships between them. Now, if the *relative phase* between the three mass states does not change with time, then the result of the three linear superpositions, defining the three flavour states, also does not change with time.

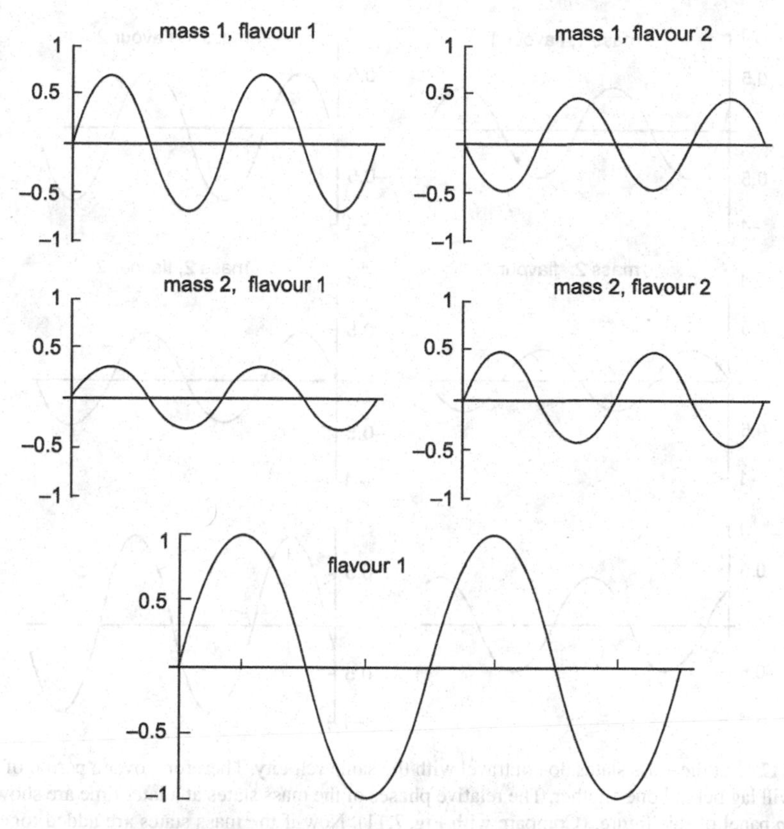

Fig. 7.11 Neutrino flavour states as superposition of mass eigenstates. The top panel shows the mass states that have to be added to obtain the flavour states. The amplitudes, and the relative phases, of the mass states have been adjusted such that when the mass states shown on the *left side* are added then one obtains the amplitude of the flavour 1 state. As may be seen from the *bottom* panel, the amplitude of the flavour 1 state thus obtained oscillates between +1 to −1. On the other hand, when the mass states shown on the right are added to obtain the flavour state 2, they cancel out precisely. Therefore, if the relative phases of the mass states *do not change with time*, then we will have a pure flavour 1 state *at all times*

However, the important thing to appreciate is that different mass eigenstates move with slightly different velocities, precisely because they have different masses. As a consequence of this, *the relative phases between the three mass states will change with time.* Therefore, a linear superposition of the mass states which initially corresponded to our prescription for, say, an electron neutrino, will no longer be so at all times. This is just an interference phenomenon (see Figs. 7.11 and 7.12).

In optics, when we talk of *constructive* and *destructive* interference of two waves, what we have in mind are two different *path lengths* which differ by an integral multiple of $\lambda/2$. As the path length changes we get an alternating band of bright and dark fringes. In the present context, we are talking about the relative phases between the mass states changing with time because *one is*

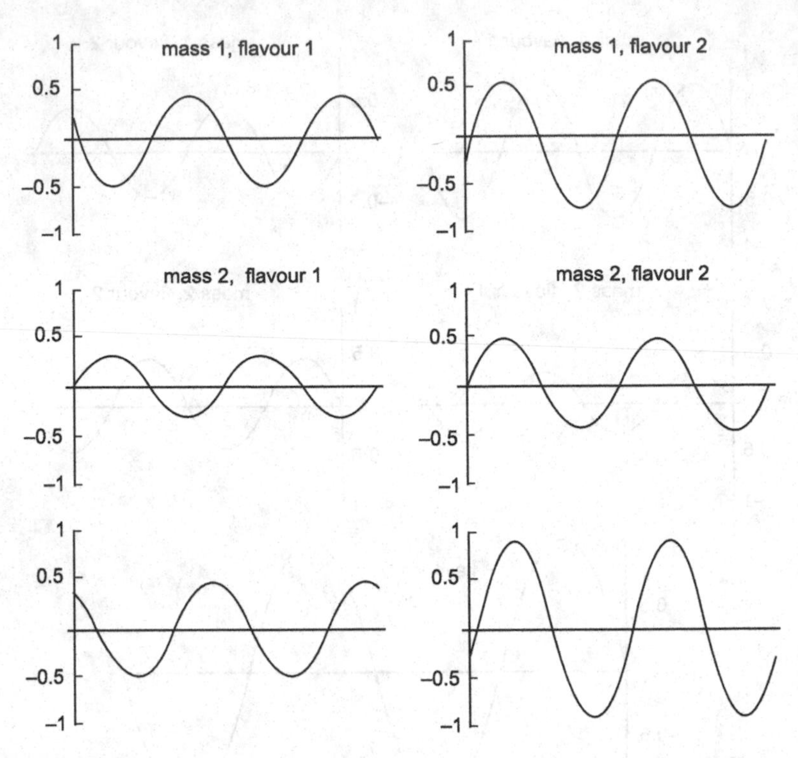

Fig. 7.12 But the mass states do not travel with the same velocity. Therefore, over a period of time they will lag behind one another. The relative phases of the mass states at a later time are shown in the *top* panel of this figure. (Compare with Fig. 7.11). Now if the mass states are added together, the flavour states one obtains are shown in the lower panel. Notice that we no longer have just the flavour 1 state. What one has is an *admixture* of the two flavour states. At a still later time, flavour 1 will be completely absent, and one will have a pure flavour 2 state. This is the phenomenon of neutrino oscillations

lagging behind the other. We shall make these remarks more comprehensible by writing a few simple equations. But before that, let us look at a couple of phenomena which are classical analogues of the quantum oscillation phenomenon we are discussing.

Coupled Pendulums

Before proceeding further, let us look at couple of classical analogies. I am sure you are familiar with coupled pendulums. Try to set up the simple experiment shown in Fig. 7.13. Erect two rigid stands or something equivalent. Tie a string between them. Now suspend two pendulums from this commons string as shown in the figure. Stand right in front of pendulum 1 (P_1) and pull it towards you by holding the bob of the pendulum between your thumb and the index finger. Carefully release it. The

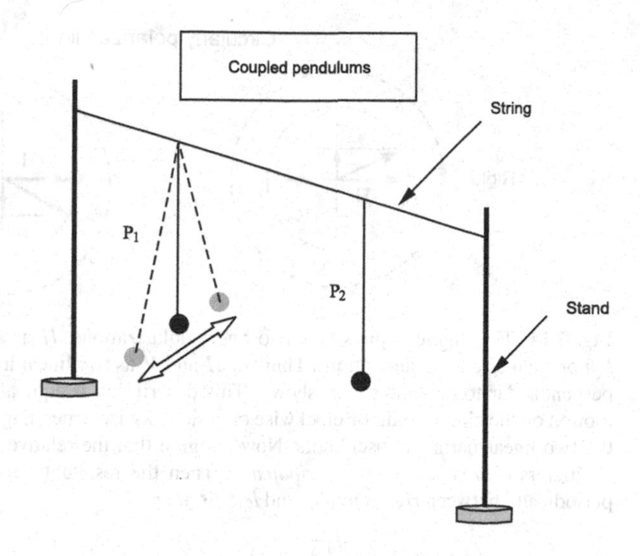

Fig. 7.13 Since the two pendulums are 'coupled' by the string from which they are suspended, the oscillation of one of them will induce oscillation in the other. If the two pendulums are of identical length and tension, then the first one will completely stop, while the second one will oscillate with the maximum amplitude

pendulum will start oscillating. After a while, its amplitude will decrease but the second pendulum (P_2) will now start oscillating. Soon P_1 will stop and P_2 will be oscillating with maximum amplitude. And the whole thing will repeat again. Try it! Increase the number of pendulums. Make them of different lengths and tension. You are guaranteed great fun!

Let us now say that the oscillation of the pendulum P_1 corresponds to the propagation of the electron neutrino, and the oscillation of the second pendulum P_2 corresponds to the propagation of, say, the muon neutrino. The common string from which the two are suspended represents a coupling or *mixing* between the two states. The quantum analogue of this classical system would be the following. Initially, as the neutrino starts propagating, it can be identified as a pure electron neutrino. Some time later—when both the pendulums are excited—what we have is an admixture of an electron neutrino and a muon neutrino. After some time, the propagating neutrino would be identified as a pure muon neutrino. And this *oscillation* between the two flavours will continue.

Is this not a marvellous example? This classical analogy was first pointed out by independently by Mikheyev and Smirnov (1986) in Russia and Steven Weinberg (1987) in U.S.A. By the way, Steven Weinberg, along with Abdus Salam and Shendon Glashow, formulated the famous Electro-Weak Unified Theory for which they were jointly awarded the Nobel Prize for Physics.

Circular Polarization of Light

Another analogy, this time from optics, might make neutrino flavour oscillations easier to visualize. The state of polarization of light can be described either in terms

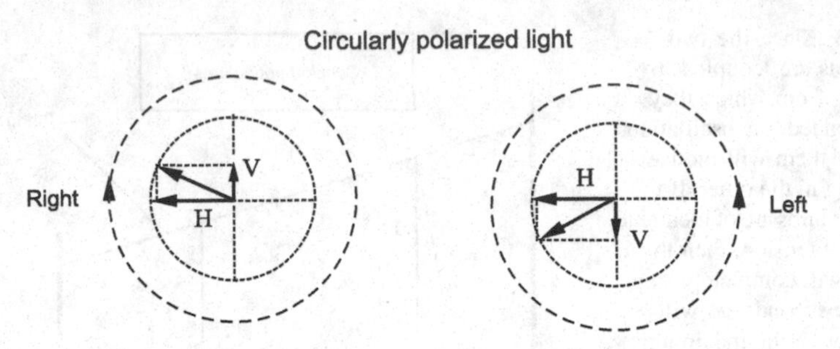

Fig. 7.14 This figure depicts how two linear polarizations, H and V, can be combined to give *left* or *right* circular polarizations. Think of H and V as two linear harmonic oscillators oscillating perpendicular to one another as shown. This description is equivalent to a circular motion. The motion on the circle would be clockwise or anticlockwise depending on the relative phase between the two linear harmonic oscillators. Now imagine that the relative phase between the two linear oscillators *changes with time periodically.* Then the resultant circular motion will also change periodically between *right* circular and *left* circular

of *linear* polarization or *circular* polarization. The term *polarization* refers to the direction of electric field vector. Since the electromagnetic wave is *transverse*, the electric vector must be confined to a plane perpendicular to the direction of propagation. If the electric field oscillates along a particular direction in this plane then one says that the wave is *linearly polarized* in that direction. In a circularly polarized beam of light, the electric field vector rotates in a periodic manner in a plane perpendicular to the direction of propagation. The tip of the electric field vector moves on the circumference of a circle. This rotation, with respect to the direction of propagation, can either be clockwise (right circular) or anti clockwise (left circular) (Fig. 7.14).

These two descriptions of the state of polarization in terms of linear and circular polarization are equivalent. You will recall from our discussion of Zeeman effect in Chap. 6 that linear harmonic motion can be represented as a superposition of two circular motions in opposite directions. Similarly, circular motion can be represented as a superposition of two linear harmonic motions, perpendicular to each other. The resultant circular motion will be either clockwise or anticlockwise depending upon the phase difference between the two perpendicular linear harmonic motions. Therefore, a linear superposition of two orthogonal linear polarizations with a suitable phase relation between the two can produce either a right circular polarization or a left circular. Now imagine that the *relative phases of the two linear oscillations changes periodically*. Then, given enough time, what was initially right circularly polarized light will become left circular. Indeed, it will oscillate periodically between right circular and left circular. This transformation will not occur if the relative phase of the two linear oscillators does *not* change with time.

Quantum Oscillations

Fortified with these two familiar examples, let us outline a slightly more formal description of neutrino oscillations. Let us denote the mass eigenstates by $|\nu_1\rangle$, $|\nu_2\rangle$ etc. and the flavour eigenstates (ν_e, ν_μ, ν_τ) as $|\nu_\alpha\rangle$, $|\nu_\beta\rangle$, etc. The relationship between these eigenstates is given by:

$$|\nu_\alpha\rangle = \sum_i U^*_{\alpha i}|\nu_i\rangle \\ |\nu_i\rangle = \sum_\alpha U_{\alpha i}|\nu_\alpha\rangle \tag{7.4}$$

where U is a unitary matrix. The first equation states that the flavour states (left-hand side) can be expressed as a superposition of the mass states, with a specific phase relationship between the mass states. The second equation says that the mass eigenstates can be similarly expressed as a superposition of the flavour states. For simplicity of illustration of the phenomenon of oscillation, let us restrict ourselves to just *two* types of neutrinos. In this case U is a very simple and well known 2×2 matrix,

$$U = \begin{pmatrix} \cos\theta & \sin\theta \\ -\sin\theta & \cos\theta \end{pmatrix}. \tag{7.5}$$

Here θ is known as the *mixing angle.*

Since the mass eigenstates are the normal modes of the system, their propagation can be described by *plane wave* solutions of the form

$$|\nu_i(t)\rangle = e^{-\frac{i}{\hbar}(E_i t - p_i x)}|\nu_i(0)\rangle \tag{7.6}$$

where E_i is the energy, p_i the momentum and x the position of the particle at time t. This equation describes the time evolution of the mass eigenstate. You may be more familiar with the phase factor of a plane wave written as $e^{i(\omega t - kx)}$. It is the same thing. Instead of ω we have used the energy E, related by $E = \hbar\omega$. Similarly, the momentum p and wave vector k are related by $p = \hbar k$. The time evolution of a flavour state, for example, the electron neutrino state, can now be explicitly written down. Using Eqs. (7.4) and (7.5) we can write the electron neutrino state as a superposition of the two mass states as follows:

$$|\nu_e(t)\rangle = \cos\theta\,|\nu_1(t)\rangle + \sin\theta\,|\nu_2(t)\rangle \tag{7.7}$$

Using the time evolution given in Eq. (7.6) this can be written as:

$$|\nu_e\,(t)\rangle = \cos\theta\; e^{-\frac{i}{\hbar}E_1 t}|\nu_1(0)\rangle + \sin\theta\; e^{-\frac{i}{\hbar}\,E_2 t}|\nu_2(0)\rangle \tag{7.8}$$

This tells us how the electron neutrino state, expressed as a linear superposition of the mass states $|\nu_1\rangle$, $|\nu_2\rangle$, changes with time. We can now ask what is the *amplitude* of an electron neutrino remaining an electron neutrino after travelling for a time t. According to the prescription of quantum mechanics, this is given by:

$$\langle \nu_e(t) \mid \nu_e(t) \rangle = \cos^2 \theta e^{-\frac{i}{\hbar} E_1 t} + \sin^2 \theta e^{-\frac{i}{\hbar} E_2 t} \tag{7.9}$$

If you are not familiar with quantum mechanics, just accept the above equation.

We can now ask the following. What is the *probability* that a flavour state, which was an electron neutrino at time $t = 0$, remains an electron neutrino at a later time? In quantum mechanics, the probability is given by the *square of the modulus of the amplitude*. Hence, the desired probability is given by:

$$\boxed{|\langle \nu_e(t) \mid \nu_e(t) \rangle|^2 = 1 - \sin^2 2\theta \sin^2 \left[\tfrac{1}{2\hbar}(E_2 - E_1)t\right].} \tag{7.10}$$

It is quite simple to derive this from Eq. (7.9), but we shall not attempt to do it here. (If you would like to convince yourself of this, work it out. Recall that $e^{i\theta} = \cos\theta + i \sin\theta$, $|x + iy|^2 = (x^2 + y^2)$ and $2\cos\theta \sin\theta = \sin 2\theta$. Using these formulae, you should easily be able to derive Eq. (7.10) from Eq. (7.9).)

The expression for the probability given in equation (7.10) is the result we wished to establish. We see that the probability is an *oscillatory function of time*. The frequency of oscillation is $(E_2 - E_1)/2\hbar$. The important thing to note is that the probability is not *unity* at all times. What does it mean to say that the probability of the electron neutrino flavour state is *less than unity* at some later time? After all, this probability was *unity* at $t = 0$; we started out with an electron neutrino flavour state. Clearly, what this means is that we no longer have the original pure flavour state, but a linear superposition of the two flavour states. The probability of observing the other flavour state is obviously given by unity minus the above probability, namely:

$$|\langle \nu_x(t) \mid \nu_e(t) \rangle|^2 = \sin^2 2\theta \;\; \sin^2 \left[\tfrac{1}{2\hbar}(E_2 - E_1)t\right]. \tag{7.11}$$

At an even later time, the probability of the original electron neutrino flavour state will be a minimum, while the probability of the second flavour state $|\nu_x(t)\rangle$ will be the maximum. The maximum probability of conversion is equal to $\sin^2 2\theta$ where θ is the *mixing angle* (see Eqs. (7.5) and (7.10)). When $\sin^2 2\theta = 1$, then the probability oscillates from 100 percent for the first flavour to 100 percent for the second flavour. This phenomenon of oscillation between two flavour states is shown in Fig. 7.15.

Instead of expressing the probability as an oscillatory function of *time*, we can also express it as an oscillatory function of the *distance travelled*. The distance travelled in a given time is R = velocity × time. Let us rewrite Eq. (7.10) in terms of the distance travelled. Let us assume that our neutrinos are ultrarelativistic with speed very close to the speed of light. In this limit, we can approximate the expression relating the energy and momentum given by the Special Theory of Relativity:

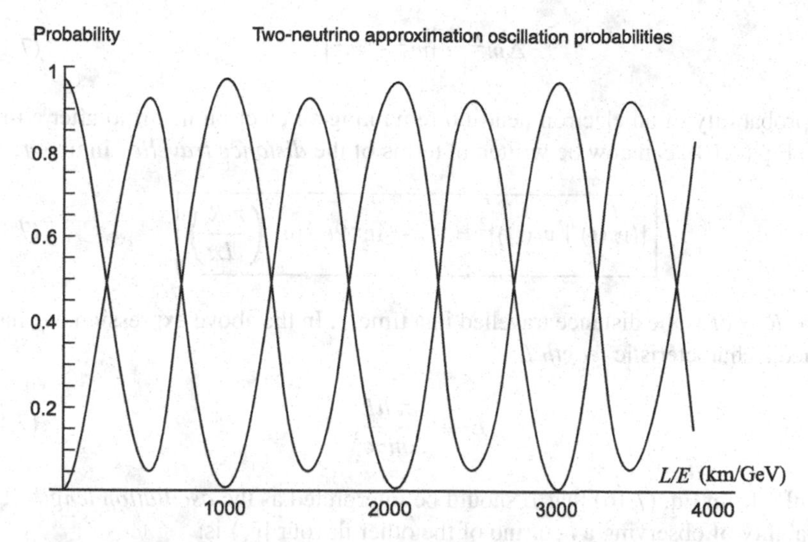

Fig. 7.15 The probability of observing the two flavours as a function of time, t, or the distance travelled, R. Notice that the probability oscillates between a maximum value and a minimum value. If the mixing angle is such that $\sin^2 2\theta = 1$, then the conversion from one flavour to another is 100 percent. As may be seen from Eqs. (7.10) and (7.12), each flavour will oscillate between +1 and −1 if $\sin^2 2\theta = 1$. In the above discussion, we have assumed for simplicity that there are only two flavour states and two mass states. In reality, there are three flavour states and three mass states. The *frequency of oscillation* is determined by Δm^2 and the energy of the neutrino E (see Eq. 7.14)

$$\begin{aligned} E_i &= \sqrt{p_i^2 c^2 + m_i^2 c^4} \\ &= p_i c \left(1 + \frac{m_i^2 c^4}{p_i^2 c^2}\right)^{1/2} = p_i c \left(1 + \frac{1}{2}\frac{m_i^2 c^4}{p_i^2 c^2} + ...\right) \\ &\cong p_i c + \frac{m_i^2 c^4}{2 p_i c} \cong E + \frac{m_i^2 c^4}{2E} \end{aligned} \tag{7.12}$$

The approximation we have made above is the following. In Special Theory of Relativity, the energy of a particle is the sum of its *kinetic energy* and *rest mass energy*. In the extreme relativistic regime, the rest mass energy is negligible as compared to the kinetic energy, which is approximately the momentum multiplied by the velocity of light. We can therefore expand the expression for the energy in powers of the ratio $(mc^2/pc) \ll 1$ and keep only the lowest-order term.

Let us use the approximation given in Eq. (7.13) in the expression for the probability given in Eq. (7.10). The energy difference $E_2 - E_1$ can now be approximated as follows:

$$E_2 - E_1 = \frac{(m_2^2 - m_1^2)c^4}{2E} \equiv \pm\frac{\Delta m^2 c^4}{2E}, \tag{7.13}$$

where,

$$\Delta m^2 \equiv \left| m_1^2 - m_2^2 \right| . \tag{7.14}$$

The probability of an electron neutrino remaining an electron neutrino after a time t, see Eq. (7.10), can now be written in terms of the *distance travelled* in time t:

$$\boxed{|\langle \nu_e(t) \mid \nu_e(t) \rangle|^2 = 1 - \sin^2 2\theta \sin^2 \left(\frac{\pi R}{L} \right)} \tag{7.15}$$

where $R = ct$ is the distance travelled in a time, t. In the above expression, we have defined a characteristic *length* L:

$$L \equiv \frac{4\pi \hbar E}{\Delta m^2 c^3} \tag{7.16}$$

We infer from Eq. (7.16) that L should be interpreted as the *oscillation length.* The probability of observing a neutrino of the other flavour $|\nu_x\rangle$ is:

$$\boxed{|\langle \nu_x(t) \mid \nu_e(t) \rangle|^2 = \sin^2 2\theta \sin^2 \left(\frac{\pi R}{L} \right)} \tag{7.17}$$

This is the phenomenon of neutrino oscillation. Since we have assumed that our neutrinos are propagating in vacuum, L, should properly be called the *vacuum oscillation* length. Note that this length is determined by two things: E, *the energy of the neutrino* and Δm^2, *the difference between the square of the masses of the eigenstates.* Clearly, if all the neutrino states have zero mass, or if their masses are the same, then there cannot be any oscillation; the oscillation length is *infinity.*

Neutrino Oscillations in Matter

How does the presence of matter change all this? After all, our solar neutrinos have to travel the first million kilometres inside the Sun. The effect of matter on the oscillations was investigated by two physicists in Russia, namely Mikheyev and Smirnov in a series of very important papers in 1986 (these are the same two persons who suggested the classical analogy of coupled pendulums to understand the oscillation phenomenon). They built upon some fundamental results obtained by Wolfenstein in 1978. Hence the effect of matter on the oscillation is known as the **MSW effect**, named after the three discoverers Mikheyev, Smirnov and Wolfenstein. We shall not pause to discuss this fascinating phenomenon here, but merely make the following remark in passing. It turns out that if the neutrino is travelling in a medium in which the *density is varying*, then the effect of the matter on the oscillation can be dramatic. There can be a *resonant* conversion of one flavour state into another. Recall that inside the Sun the density decreases from approximately 150 g/cm^3 near the centre

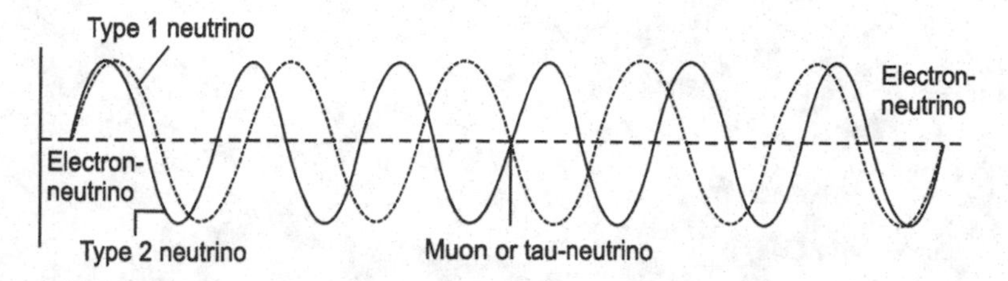

Fig. 7.16 This figure illustrates how the flavour state oscillates between an electron neutrino and a muon (or tau neutrino). Initially, the mass states 1 and 2 are *in phase*. Let us define their superposition as the electron neutrino. After propagating for a certain distance, the two mass states are precisely *out of phase*. Let us define this as the muon neutrino. Clearly, the flavour state will oscillate between electron and muon neutrino states

to less than the density of air in the outer region. The relevance of the MSW effect to understand the solar neutrino puzzle was first pointed out by Hans Bethe in 1986. He pointed out that by the time an electron neutrino leaves the Sun its flavour could have changed completely due to the MSW effect.

Let us conclude the discussion of neutrino oscillations by reiterating that the oscillation between different flavours is a simple consequence of the *mass eigenstates propagating with slightly different speeds*. This causes the *relative phase between them to vary with time* as they propagate. Therefore the states which are superpositions of these mass states will naturally oscillate with time (Recall our discussion of circular polarization of light). It is as simple as that. For this to happen, at least some of the neutrinos must have rest masses which are different from each other. It is the mass that makes the neutrino eigenstate of a definite momentum travel with different speeds.

Neutrino oscillation during propagation is graphically represented in Fig. 7.16. For simplicity, we have considered just two flavours, say, electron neutrino and muon neutrino. Let the two mass eigenstates be represented by the *solid* and *dashed* sine waves. The changing phase relationship between them is indicated by the two sine waves having different wavelengths. Let us define the electron neutrino as a linear combination of the two sine waves added *in phase*. After travelling some distance, they will no longer be *in phase*. So the linear superposition will no longer be a pure electron neutrino. At some stage, the two waves will be 180° out of phase, making it a pure muon neutrino. So the probability of the electron neutrino retaining its original identity will oscillate. We see from (7.16) that the maximum probability of conversion is equal to $\sin^2 2\theta$, where θ is the mixing angle. The frequency of oscillation is determined by the energy of the neutrino and Δm^2.

Epilogue

It took 80 years, but in the end Eddington was proved right. Let us listen to Eddington once again:

> To my mind the "existence" of helium is the best evidence we could desire of the possibility of the "formation" of helium. The four protons and two electrons constituting its nucleus must have been assembled at some time and place; and why not in the stars?
>
> ... *I am aware that many critics consider the conditions in the stars not sufficiently extreme to bring about the transmutations–the stars are not hot enough. The critics lay themselves open to an obvious retort; we tell them to go and find "a hotter place".*

It is sobering to note that these words were uttered in 1926. The helium nucleus does not contain *four protons and two electrons*, as Eddington had said. But the neutron had not yet been discovered when Eddington made that comment. We now know that the helium nucleus consists of two protons and two *neutrons*. Two neutrinos must be emitted in the process of transforming two protons into two neutrons. The observed luminosity of the Sun implies that roughly six hundred million metric tons of hydrogen is being converted to helium every second. If so, the Sun must emit 2×10^{38} electron neutrinos every second. Detecting these neutrinos would vindicate Eddington's prescient assertion. But, alas, the neutrinos are elusive! More than that, they masquerade as neutrinos of different types—to avoid detection. But physicists are very clever, and persistent. They found ways of detecting all the solar neutrinos, despite their attempt to conceal themselves. In the process, physicists have proved that the astronomer's model of the Sun, and why it shines, is spectacularly correct. But this has come at a price. The Standard Model of elementary particles needs to be modified.

Can Stars Find Peace? A Sneak Preview

This first volume of this series was devoted to the question 'What Are the Stars and why Are They As They Are?' We discussed the stability of the stars and why they shine. Along the way, we discussed many other things, some of them quite recent

G. Srinivasan, *What are the Stars?* Undergraduate Lecture Notes in Physics,
DOI: 10.1007/978-3-642-45302-1, © Springer-Verlag Berlin Heidelberg 2014

developments. For, example, we discussed at length how the spontaneous oscillations of the surface of the Sun was used to infer the internal conditions with unprecedented accuracy. The last chapter was devoted to how clinching evidence was finally found to prove Eddington's conjecture that the Sun shines because it is converting hydrogen to helium at its centre.

This is only the first act of the drama. Having created a substantial amount of helium in its core, the Sun will go on to fuse helium into carbon, carbon into oxygen, and so on. Each stage of transmutation of elements is another 'Act' of the story. We shall discuss the various acts in the life history of stars in the next volume of the series.

But a more interesting question is this: how will the story end? What will happen to a star when the nuclear reactions cease? This can happen for two reasons. Either the star runs out of fuel, or the nuclear reactor at the centre switches off because the core is not hot enough. What will then happen to the star depends on how massive it is. Low mass stars like the Sun will collapse and end their lives peacefully when they reach a density of the order of 10^6 g cm^{-3}. Such end states of stars are known as *white dwarfs*. When they cool sufficiently, they will crystallize, and become gigantic *diamonds in the sky* and they will live for ever! Their ultimate peace is guaranteed by quantum physics. A discussion of these ideas will be one of the main themes of the next volume entitled ***Can Stars Find Peace?***[1]

While quantum physics guarantees that white dwarfs will be stable forever, there is a fine print. In 1930, Subrahmanyan Chandrasekhar made the sensational discovery that white dwarfs cannot be stable if their mass exceeded 1.4 $M_\odot$. This has come to be known as the *Chandrasekhar limit.* This raises the following question. Can massive stars find peace? Or, are they doomed? The answer to this question became clear in the 1980s. Stars with mass roughly in the range 10–25 $M_\odot$ will find ultimate peace as *neutron stars*. Such stars will have radii of about 10 km and density of about 10^{14} g cm^{-3}. When the cores of massive stars collapse to form neutron stars, the gravitational potential energy released will be so enormous that the envelope of the star will be blown up in a spectacular explosion. Such stellar explosions are known as *supernovae.* Even more massive stars cannot find peace as neutron stars. An appeal to quantum physics cannot be made to save them. They are doomed. They will collapse to become *black holes*.

To know more about *white dwarfs*, the *Chandrasekhar limit* and *supernovae*, you have to wait for the next volume!

[1] The Life and Death of Stars (in the Springer edition)

Suggested Reading

1. A S Eddington, *Stars and Atoms*, Oxford University Press, 1927.
   This book, based on some of Eddington's public lectures is a masterpiece. You may find it difficult to access this book, but it would be well worth your while to hunt for it.
2. A S Eddington, *The Internal Constitution of the Stars*, Cambridge University Press, 1926.
   Eddington explained his theory of the stars in this famous book. This has been reprinted many times. The 1988 edition has a wonderful foreword by S Chandrasekhar. This is an advanced book, but it is worth looking at the first chapter in which Eddington surveys the problem.
3. George Gamow, *The Birth and Death of the Sun*, Mentor Book, 1940.
   Another masterpiece. Gamow wrote many popular books in which he explained in simple terms some of the latest developments in physics.
4. R Kippenhahn, *Discovering the Secrets of the Sun*, John Wiley & Sons, 1994.
5. G Venkataraman, *Saha and His Formula*, Universities Press (India), 1995
6. G Venkataraman, *The Big and the Small: The Journey into the Microcosm*, Universities Press (India), 2001.
   If you wish to know more about the elementary particles that are the constituents of matter, then you will find this book most enjoyable. It narrates the story from the discovery of the electron to the discovery of quarks. There is a nice discussion of neutrinos in this book.

G. Srinivasan, *What are the Stars?* Undergraduate Lecture Notes in Physics,
DOI: 10.1007/978-3-642-45302-1,

Index

G. Srinivasan, *What are the Stars?* Undergraduate Lecture Notes in Physics,
DOI: 10.1007/978-3-642-45302-1, © Springer-Verlag Berlin Heidelberg 2014